Edexcel Award in Algebra

Diane Oliver

ALWAYS LEARNING

PEARSON

Published by Pearson Education Limited, Edinburgh Gate, Harlow, Essex, CM20 2JE.

www.pearsonschoolsandfecolleges.co.uk

Edited by Project One Publishing Solutions, Scotland
Typeset and illustrated by Tech-Set Ltd, Gateshead

First published 2013

17 16 15 14 13
10 9 8 7 6 5 4 3 2 1

British Library Cataloguing in Publication Data
A catalogue record for this book is available from the British Library

ISBN 978 1 446 90323 0

Printed in Slovakia by Neografia

Acknowledgements
Every effort has been made to contact copyright holders of material reproduced in this book. Any omissions will be rectified in subsequent printings if notice is given to the publishers.

Notices

The AS Links provide references to course books as follows:

C1 Edexcel AS and A Level Modular Mathematics Core Mathematics 1
ISBN 978 0 43551 910 0

C2 Edexcel AS and A Level Modular Mathematics Core Mathematics 2
ISBN 978 0 43551 911 7

D1 Edexcel AS and A Level Modular Mathematics Decision Mathematics 1
ISBN 978 1 84690 893 4

M1 Edexcel AS and A Level Modular Mathematics Mechanics Mathematics 1
ISBN 978 0 43551 916 2

Contents

Self-assessment chart iv

Chapter 1 Algebraic manipulation **1**
1.1 Expanding two brackets 1
1.2 Factorising expressions 2
1.3 Using index laws 4
1.4 Algebraic fractions 6
1.5 Completing the square 8

Chapter 2 Formulae **11**
2.1 Substitution 11
2.2 Changing the subject of a formula 12

Chapter 3 Surds **14**
3.1 Surds 14
3.2 Rationalising the denominator 15

Chapter 4 Quadratic equations **18**
4.1 Solving by factorisation 18
4.2 Solving by completing the square 20
4.3 Solving by using the formula 21

Chapter 5 Roots of quadratic equations **24**
5.1 The role of the discriminant 24
5.2 The sum and product of the roots of a quadratic equation 26

Chapter 6 Simultaneous equations **28**
6.1 Solving simultaneous linear equations using elimination 28
6.2 Solving simultaneous linear equations using substitution 29
6.3 Solving simultaneous equations where one is quadratic 30

Chapter 7 Arithmetic series **33**
7.1 General (nth) term of arithmetic series 33
7.2 The sum of an arithmetic series 35

Chapter 8 Coordinate geometry **40**
8.1 The equation of a line 40
8.2 Parallel and perpendicular lines 43

Chapter 9 Graphs of functions **47**
9.1 Recognising graphs 47
9.2 Drawing and using graphs 49
9.3 Sketching graphs 52
9.4 Graphs of circles 57

Chapter 10 Inequalities **64**
10.1 Solving linear inequalities 64
10.2 Solving quadratic inequalities 65
10.3 Representing linear inequalities on a graph 67

Chapter 11 Distance–time and speed–time graphs **72**
11.1 Distance–time graphs 72
11.2 Speed–time graphs 73

Chapter 12 Direct and inverse proportion **78**
12.1 Direct proportion 78
12.2 Inverse proportion 81

Chapter 13 Transformations of functions **86**
13.1 Applying the transformations $y = f(x) \pm a$ and $y = f(x \pm a)$ to the graph of $y = f(x)$ 86
13.2 Applying the transformations $y = f(\pm ax)$ and $y = \pm af(x)$ to the graph of $y = f(x)$ 88

Chapter 14 Area under a curve **93**
14.1 The trapezium rule 93

Practice Paper **100**

Answers **110**

Name .... Class ....

Self-assessment chart

	Needs more practice	Almost there	I'm proficient!	Notes
Chapter 1 Algebraic manipulation				
1.1 Expanding two brackets	☐	☐	☐	
1.2 Factorising expressions	☐	☐	☐	
1.3 Using index laws	☐	☐	☐	
1.4 Algebraic fractions	☐	☐	☐	
1.5 Completing the square	☐	☐	☐	
Chapter 2 Formulae				
2.1 Substitution	☐	☐	☐	
2.2 Changing the subject of a formula	☐	☐	☐	
Chapter 3 Surds				
3.1 Surds	☐	☐	☐	
3.2 Rationalising the denominator	☐	☐	☐	
Chapter 4 Quadratic equations				
4.1 Solving by factorisation	☐	☐	☐	
4.2 Solving by completing the square				
4.3 Solving by using the formula	☐	☐	☐	
Chapter 5 Roots of quadratic equations				
5.1 The role of the discriminant	☐	☐	☐	
5.2 The sum and product of the roots of a quadratic equation	☐	☐	☐	
Chapter 6 Simultaneous equations				
6.1 Solving simultaneous linear equations using elimination	☐	☐	☐	
6.2 Solving simultaneous linear equations using substitution	☐	☐	☐	
6.3 Solving simultaneous equations where one is quadratic	☐	☐	☐	
Chapter 7 Arithmetic series				
7.1 General (nth) term of arithmetic series	☐	☐	☐	
7.2 The sum of an arithmetic series	☐	☐	☐	
Chapter 8 Coordinate geometry				
8.1 The equation of a line	☐	☐	☐	
8.2 Parallel and perpendicular lines	☐	☐	☐	
Chapter 9 Graphs of functions				
9.1 Recognising graphs	☐	☐	☐	
9.2 Drawing and using graphs	☐	☐	☐	
9.3 Sketching graphs	☐	☐	☐	
9.4 Graphs of circles	☐	☐	☐	
Chapter 10 Inequalities				
10.1 Solving linear inequalities	☐	☐	☐	
10.2 Solving quadratic inequalities	☐	☐	☐	
10.3 Representing linear inequalities on a graph	☐	☐	☐	
Chapter 11 Distance–time and speed–time graphs				
11.1 Distance–time graphs	☐	☐	☐	
11.2 Speed–time graphs	☐	☐	☐	
Chapter 12 Direct and inverse proportion				
12.1 Direct proportion	☐	☐	☐	
12.2 Inverse proportion	☐	☐	☐	
Chapter 13 Transformations of functions				
13.1 Applying the transformations $y = f(x) \pm a$ and $y = f(x \pm a)$ to the graph of $y = f(x)$	☐	☐	☐	
13.2 Applying the transformations $y = f(\pm ax)$ and $y = \pm af(x)$ to the graph of $y = f(x)$	☐	☐	☐	
Chapter 14 Area under a curve				
14.1 The trapezium rule	☐	☐	☐	

Needs more practice ☐ Almost there ☐ I'm proficient! ☐

1.1 Expanding two brackets

By the end of this section you will know how to:

* Multiply out two linear expressions

Key points

* When you expand one set of brackets you must multiply everything inside the bracket by what is outside.
* When you expand two linear expressions, each with two terms of the form $ax + b$, where $a \neq 0$ and $b \neq 0$, you create four terms. Two of these can usually be simplified by collecting like terms.

Guided

1 Expand and simplify

a $3x(2x - 5)$

$6x^{\dots} - \dots$

Hint
Multiplying two numbers of the same sign gives a positive answer.
Multiplying two numbers of different signs gives a negative answer.

b $(x + 3)(x + 2)$

Using the grid method

$(x + 3)(x + 2)$

$= x^2 + 2x + \dots + \dots$

$= x^2 + \dots + \dots$

$\times$	x	$+2$
x	x^2	$+2x$
$+3$	....	....

c $(x - 5)(2x + 3)$

$= 2x^2 - 10x + \dots - \dots$

$= 2x^2 \dots$

d $(2x - 5y)(3x - 4y)$

$= 6x^2 - \dots - \dots + \dots$

$= 6x^2 \dots$

Practice

2 Expand and simplify

a $2x(x + 4)$

b $6x(3x - 5)$

c $5x(2x + 2y)$

d $(x + 4)(x + 5)$

e $(x + 7)(x + 3)$

f $(x + 7)(x - 2)$

g $(x + 5)(x - 5)$

h $(2x + 3)(x - 1)$

i $(3x - 2)(2x + 1)$

j $(5x - 3)(2x - 5)$

k $(3x - 2)(7 + 4x)$

l $(3x + 4y)(5y + 6x)$

m $(5x + 2y)(7x - 3)$

n $(3x - 8)(2x - 3y)$

Needs more practice ☐ Almost there ☐ I'm proficient! ☐

1.2 Factorising expressions

By the end of this section you will know how to:

* Factorise expressions by taking out common factors
* Factorise quadratic expressions

AS LINKS
C1: 1.4 Factorising expressions; 1.5 Factorising quadratic expressions

Key points

* **Factorising** an expression is the opposite of expanding the brackets.
* A **quadratic expression** is in the form $ax^2 + bx + c$, where $a \neq 0$.
* To factorise a quadratic equation find two numbers whose sum is b and whose product is ac.
* An expression in the form $x^2 - y^2$ is called the **difference of two squares**. It factorises to $(x - y)(x + y)$

Guided

1 Factorise

Hint
Take the highest common factor outside the bracket

a $15x^2y^3 + 9x^4y$

$= 3x^2y(5y^2 + \ldots\ldots)$

b $4x^2 - 25y^2$

$= (2x - 5y)(\ldots\ldots + \ldots\ldots)$

c $x^2 + 3x - 10$

$b = \ldots\ldots$, $ac = -10$

Two numbers are 5 and -2

$x^2 + 3x - 10 = x^2 + 5x - 2x - 10$

$= x(x + 5) - 2(x + \ldots\ldots)$

$= (x + 5)(x - \ldots\ldots)$

d $6x^2 - 11x - 10$

$b = \ldots\ldots$, $ac = \ldots\ldots$

Two numbers are -15 and $\ldots\ldots$

$6x^2 - 11x - 10 = 6x^2 - 15x + \ldots\ldots - 10$

$= 3x(\ldots\ldots) + 2(\ldots\ldots)$

$= (3x + 2)(\ldots\ldots)$

Practice

2 Factorise

a $6x^4y^3 - 10x^3y^4$

b $21a^3b^5 + 35a^5b^2$

c $25x^2y^2 - 10x^3y^2 + 15x^2y^3$

3 Factorise

a $x^2 + 7x + 12$

b $x^2 + 5x - 14$

c $x^2 - 11x + 30$

d $x^2 - 5x - 24$

e $x^2 - 7x - 18$

f $x^2 + x - 20$

g $x^2 - 3x - 40$

h $x^2 + 3x - 28$

4 Factorise

a $36x^2 - 49y^2$

b $4x^2 - 81y^2$

c $18a^2 - 200b^2c^2$

5 Factorise

a $2x^2 + x - 3$

b $6x^2 + 17x + 5$

c $12x^2 - 38x + 20$

d $2x^2 + 7x + 3$

e $9x^2 - 15x + 4$

f $10x^2 + 21x + 9$

Needs more practice ☐ Almost there ☐ I'm proficient! ☐

1.3 Using index laws

AS LINKS
C1: 1.2 The rules of indices

By the end of this section you will know how to:

* Use index laws with fractional and negative indices

Key points

* $a^m \times a^n = a^{m+n}$
* $(a^m)^n = a^{mn}$
* $a^{\frac{1}{n}} = \sqrt[n]{a}$ i.e. the nth root of a
* $a^{-m} = \dfrac{1}{a^m}$
* $\dfrac{a^m}{a^n} = a^{m-n}$
* $a^0 = 1$
* $a^{\frac{m}{n}} = \sqrt[n]{a^m} = (\sqrt[n]{a})^m$
* The square root of a number produces two solutions, e.g. $\sqrt{16} = \pm 4$.

Guided

1 Evaluate

a $10^0 =$ ....

Hint
Any value raised to the power of zero is equal to 1.

b $9^{\frac{1}{2}}$

$= \sqrt{.......}$

$=$ ....

c $27^{\frac{2}{3}}$

$= \sqrt[3]{.......}^{\,2}$

$= 3^{....} =$ ....

d 4^{-2}

$= \dfrac{1}{.......}$

$= \dfrac{1}{.......}$

e $\dfrac{6x^5}{2x^2}$

$= 3x^{....}$

f $\dfrac{x^3 \times x^5}{x^4}$

$= \dfrac{x^{....}}{x^{....}}$

$= x^{....}$

g $\dfrac{x^5}{x^2 \times x^{\frac{1}{2}}}$

$= \dfrac{x^{....}}{x^{....}}$

$= x^{....}$

h $\dfrac{3x^{\frac{3}{2}} \times 4x^{\frac{1}{2}}}{(2x^2)^3}$

$= \dfrac{......x^{....}}{......x^{....}}$

$=$ ....

Practice

2 Evaluate

a 14^0

b 3^0

c x^0

3 Evaluate

a $49^{\frac{1}{2}}$

b $64^{\frac{1}{3}}$

c $125^{\frac{1}{3}}$

d $16^{\frac{1}{4}}$

4 Evaluate

a $25^{\frac{3}{2}}$ **b** $8^{\frac{5}{3}}$ **c** $49^{\frac{3}{2}}$ **d** $16^{\frac{3}{4}}$

5 Evaluate

a 5^{-2} **b** 4^{-3} **c** 2^{-5} **d** 6^{-2}

6 Simplify

a $\frac{3x^2 \times x^3}{2x^2}$ **b** $\frac{10x^5}{2x^2 \times x}$ **c** $\frac{3x \times 2x^3}{2x^3}$ **d** $\frac{7x^3y^2}{14x^5y}$

e $\frac{y^2}{y^{\frac{1}{2}} \times y}$ **f** $\frac{c^{\frac{1}{2}}}{c^2 \times c^{\frac{3}{2}}}$ **g** $\frac{(2x^2)^3}{4x^0}$ **h** $\frac{x^{\frac{1}{2}} \times x^{\frac{3}{2}}}{x^{-2} \times x^3}$

7 Evaluate

a $4^{-\frac{1}{2}}$ **b** $27^{-\frac{2}{3}}$ **c** $9^{-\frac{1}{2}} \times 2^3$

d $16^{\frac{1}{4}} \times 2^{-3}$ **e** $\left(\frac{9}{16}\right)^{-\frac{1}{2}}$ **f** $\left(\frac{27}{64}\right)^{-\frac{2}{3}}$

Needs more practice ☐ Almost there ☐ I'm proficient! ☐

1.4 Algebraic fractions

By the end of this section you will know how to:

* Simplify algebraic fractions
* Add and subtract algebraic fractions

AS LINKS
C2: 1.1 Simplifying algebraic fractions by division

Key points

* Use some laws of indices to simplify algebraic fractions.
* Factorise the numerator and denominator if possible.
* Any value divided by itself is 1.
* To add or subtract algebraic fractions, use the same method as for adding or subtracting numerical fractions.
* To add or subtract algebraic fractions with different denominators, find a common denominator and use this to write each fraction as an equivalent fraction.

Guided

1 Simplify the algebraic fractions.

a $\dfrac{2x^2 - 4x}{12x + 6x^2}$

$= \dfrac{2x(\dots\dots)}{6x(\dots\dots)}$

$= \dfrac{\dots\dots}{3(\dots\dots)}$

Hint
Take out the highest common factor in each term.

b $\dfrac{x^2 - 4x - 21}{2x^2 + 9x + 9}$

$= \dfrac{(\dots\dots)(\dots\dots)}{(2x + 3)(\dots\dots)}$

$= \dfrac{\dots\dots}{\dots\dots}$

c $\dfrac{x}{3} + \dfrac{2x + 1}{2}$

$= \dfrac{2x}{6} + \dfrac{6x + \dots}{6}$

$= \dfrac{8x + \dots}{\dots}$

d $\dfrac{2}{x - 3} - \dfrac{5}{x + 1}$

$= \dfrac{2x + \dots}{(x - 3)(x + 1)} - \dfrac{5x - \dots}{(x - 3)(x + 1)}$

$= \dfrac{\dots\dots}{(x - 3)(x + 1)}$

Practice

2 Simplify the algebraic fractions.

a $\dfrac{2x^2 + 4x}{x^2 - x}$

b $\dfrac{x^2 + 3x}{x^2 + 2x - 3}$

c $\dfrac{x^2 - 2x - 8}{x^2 - 4x}$

d $\dfrac{x^2 - 5x}{x^2 - 25}$

e $\dfrac{x^2 - x - 12}{x^2 - 4x}$

f $\dfrac{2x^2 + 14x}{2x^2 + 4x - 70}$

3 Simplify

a $\frac{2x}{3} + \frac{x}{5}$

b $\frac{x+1}{2} + \frac{3x}{5}$

c $\frac{2x}{7} - \frac{x}{4}$

d $\frac{3x}{4} - \frac{2x}{3}$

e $\frac{2x+1}{3} + \frac{x}{4}$

f $\frac{3x+2}{5} - \frac{x-1}{4}$

4 Simplify

a $\frac{2}{x+3} + \frac{3}{x+1}$

b $\frac{1}{x} + \frac{2}{x+3}$

c $\frac{3}{x+4} - \frac{2}{x}$

d $\frac{4}{x+1} - \frac{2}{x-1}$

e $\frac{7}{2x-3} - \frac{1}{x+1}$

f $\frac{3}{x+1} + \frac{2}{x-2}$

5 Simplify

a $\frac{9x^2 - 16}{3x^2 + 17x - 28}$

b $\frac{2x^2 - 7x - 15}{3x^2 - 17x + 10}$

c $\frac{4 - 25x^2}{10x^2 - 11x - 6}$

d $\frac{6x^2 - x - 1}{2x^2 + 7x - 4}$

Needs more practice ☐ Almost there ☐ I'm proficient! ☐

1.5 Completing the square

AS LINKS
C1: 2.3 Completing the square

By the end of this section you will know how to:

* Complete the square for quadratic expressions in the form $ax^2 + bx + c$, where $a > 0$.

Key points

* If $a \neq 1$, then factorise using a as a common factor.
* Completing the square for a quadratic expression rearranges $ax^2 + bx + c$ into the form $p(x + q)^2 + r$.

Guided

1 Complete the square for the quadratic expressions.

a $x^2 + 6x - 2$

$= (x + \ldots\ldots)^2 - 2 - 9$

$= (x + \ldots\ldots)^2 - \ldots\ldots$

Hint

In the form $(x + q)^2 + r$, $q = \frac{1}{2}b$, where b is the coefficient of x in the quadratic equation.

b $2x^2 - 5x + 1$

$= \ldots\ldots \left[x^2 - \frac{5}{2}x + \frac{1}{2}\right]$

$= \ldots\ldots [(x - \ldots\ldots)^2 + \ldots\ldots - \ldots\ldots]$

$= \ldots\ldots [(x - \ldots\ldots)^2 - \ldots\ldots]$

$= \ldots\ldots (x - \ldots\ldots)^2 - \ldots\ldots$

Practice

2 Write the following quadratic expressions in the form $(x + p)^2 + q$.

a $x^2 + 4x + 3$ **b** $x^2 - 10x - 3$ **c** $x^2 - 8x$

d $x^2 + 6x$ **e** $x^2 - 2x + 7$ **f** $x^2 + 3x - 2$

3 Write the following quadratic expressions in the form $p(x + q)^2 + r$.

a $2x^2 - 8x - 16$ **b** $4x^2 - 8x - 16$

c $3x^2 + 12x - 9$ **d** $2x^2 + 6x - 8$

4 Complete the square.

a $2x^2 + 3x + 6$

b $3x^2 - 2x$

c $5x^2 + 3x$

d $3x^2 + 5x + 3$

Don't forget!

- Expanding two linear expressions creates .... terms.
- A quadratic equation is an equation in the form ...., where $a \neq 0$.
- To factorise a quadratic equation find two numbers whose sum is .... and whose product is ....
- An expression in the form $x^2 - y^2$ is called ....
  It factorises to ....
- $a^m \times a^n =$ ....
- $\frac{a^m}{a^n} =$ ....
- $(a^m)^n =$ ....
- $a^0 =$ ....
- $a^{\frac{1}{n}} =$ ....
- $a^{\frac{m}{n}} =$ ....
- $a^{-m} =$ ....
- To simplify an algebraic fraction, firstly, factorise the .... and .... if possible.
- Any value divided by itself = ...
- To add or subtract algebraic fractions with different denominators, find a ....
  and use this to write each fraction as an .... fraction.
- Completing the square for a quadratic expression rearranges $ax^2 + bx + c$ into the form
  ....

Exam-style questions

1 **a** Expand and simplify $(3x + 2)(x - 3)$

b Factorise $12x^3y^2 + 30x^2y^5$

c Simplify $\frac{x^3 \times x^4}{x^5}$

2 **a** Simplify $\frac{x^{\frac{3}{2}}}{x \times x^{\frac{5}{2}}}$

b Factorise $x^2 + 2x - 35$

c Factorise $4x^2 - 25y^2$

3 Write the quadratic expression $x^2 + 3x - 5$ in the form $(x + p)^2 + q$, where p and q are mixed numbers.

4 Simplify $\frac{x^2 - 4}{2x^2 - x - 6}$

2.1 Substitution

By the end of this section you will know how to:

* Substitute numbers into expressions and formulae

Key points

* **Substitution** is replacing each letter with its value.
* Given the value of each letter in an expression or formula, you can work out the value of the expression or formula.

Guided

1 When $a = 8$, $b = -6$ and $c = \frac{1}{3}$, evaluate the following expressions.

a $2a + b$

$= 2 \times$ .... + ...

= .... − ....

= ....

b $a + bc$

= .... + ... × ....

= .... − ....

= ....

c $\frac{3a}{b}$

$= \frac{3 \times ....}{.......}$

= ....

d $a^c - b$

= $.......^{\frac{1}{3}}$ − ....

= .... + ...

= ....

2 Calculate the Celsius temperature, C, when the Fahrenheit temperature F is 50. Use the formula $C = \frac{5}{9}(F - 32)$.

$C = \frac{5}{9}$ of (....... − 32)

$C = \frac{5}{9}$ of ....

$C = 5 \times$ .... ÷ 9

$C =$ ....

Practice

3 When $x = \frac{1}{2}$, $y = -4$ and $z = 9$, evaluate the following expressions.

a $xy + z$

b z^x

c $y^2 + xz$

d $(y + z)^2$

e yz^2

f xyz

4 When $p = -3$, $q = 2$ and $r = 2.4$, evaluate the following expressions.

a $\frac{qr}{p}$

b $\frac{r - p}{q}$

c $q - \frac{r}{p}$

d $\frac{r}{p + q}$

5 When $d = \frac{1}{3}$, $e = -2$ and $f = \frac{1}{4}$, evaluate the following expressions, giving your answers as mixed numbers where appropriate.

a $\frac{de}{f}$ **b** def **c** $e(d + f)$ **d** $3d + \frac{e}{f}$

6 Using the formula $s = ut + \frac{1}{2}at^2$, calculate s when $t = 10$, $a = 9.8$ and $u = 12$.

Needs more practice ☐ Almost there ☐ I'm proficient! ☐

2.2 Changing the subject of a formula

By the end of this section you will know how to:

* Change the subject of a formula, where the new subject appears in the formula once.
* Change the subject of a formula, where the new subject appears in the formula twice.

Key points

* Get the terms containing the subject on one side and everything else on the other side.
* Usually, you then need to factorise the terms containing the new subject.

Guided

Make t the subject of the formulae.

1 $v = u + at$

$v - \ldots\ldots = at$

$t = \frac{v - \ldots\ldots}{\ldots\ldots}$

2 $r = 2t - \pi t$

$r = t(\ldots\ldots - \ldots\ldots)$

$t = \frac{\ldots\ldots}{\ldots\ldots}$

3 $\frac{t + r}{5} = \frac{3t}{2}$

Hint
Remove the fractions first.

$2(t + \ldots\ldots) = \ldots\ldots \times 3t$

$2t + \ldots\ldots = \ldots\ldots t$

$2r = \ldots\ldots$

$t = \frac{\ldots\ldots}{\ldots\ldots}$

4 $r = \frac{3t + 5}{t - 1}$

$r(\ldots\ldots) = 3t + 5$

$\ldots\ldots = 3t + 5$

$\ldots\ldots = 5 + r$

$\ldots\ldots(\ldots\ldots) = 5 + r$

$t = \frac{\ldots\ldots}{\ldots\ldots}$

Practice

Change the subject of each formula to the letter given in the brackets.

5 $C = \pi d$ [d]

6 $P = 2l + 2w$ [w]

7 $D = \frac{S}{T}$ [T]

8 $p = \frac{q - r}{t}$ [t]

9 $u = at - \frac{1}{2}t$ [t]

10 $V = ax + 4x$ [x]

11 $\frac{y - 7x}{2} = \frac{7 - 2y}{3}$ [y]

12 $x = \frac{2a - 1}{3 - a}$ [a]

13 $a = \frac{b - c}{d}$ [d]

14 $h = \frac{7g - 9}{2 + g}$ [g]

15 $e(9 + x) = 2e + 1$ [e]

Don't forget!

- Substitution means ……………………………………………………………………………………
- When changing the subject of a formula, get all the terms containing the new subject on one side and …………………………… on the other.

Exam-style questions

1 Make x the subject of the formula $y = \frac{2x + 3}{4 - x}$

3.1 Surds

AS LINKS
C1: 1.7 The use and manipulation of surds

By the end of this section you will know how to:

* Simplify surds
* Expand and simplify expressions involving surds

Key points

* A **surd** is the square root of a number that is not a square number, such as $\sqrt{2}$, $\sqrt{3}$, $\sqrt{5}$, etc.
* Surds are used to give the exact value for an answer.
* $\sqrt{ab} = \sqrt{a} \times \sqrt{b}$
* $\sqrt{\frac{a}{b}} = \frac{\sqrt{a}}{\sqrt{b}}$

Guided

Simplify

1 $\sqrt{50}$

$= \sqrt{25 \times .....}$

$= \sqrt{.....} \times \sqrt{.....}$

$= 5 \times \sqrt{.....}$

$= .....\sqrt{.....}$

Hint
One of the two numbers you choose at the start must be a square number.

2 $\sqrt{147} - 2\sqrt{12}$

$= \sqrt{49 \times .....} - 2\sqrt{..... \times .....}$

$= \sqrt{.....} \times \sqrt{.....} - 2\sqrt{.....} \times \sqrt{.....}$

$= ..... \times \sqrt{.....} - 2 \times ..... \times \sqrt{.....}$

$= .....\sqrt{.....}$

3 $(\sqrt{7} + \sqrt{2})(\sqrt{7} - \sqrt{2})$

$= \sqrt{.....} - \sqrt{7}\sqrt{2} + \sqrt{2}\sqrt{7} - \sqrt{4}$

$= ..... - .....$

$= .....$

Practice

4 Simplify

a $\sqrt{45}$ **b** $\sqrt{125}$ **c** $\sqrt{48}$ **d** $\sqrt{175}$

e $\sqrt{300}$ **f** $\sqrt{28}$ **g** $\sqrt{72}$ **h** $\sqrt{162}$

5 Expand and simplify

a $(\sqrt{2} + \sqrt{3})(\sqrt{2} - \sqrt{3})$ **b** $(3 + \sqrt{3})(5 - \sqrt{12})$

c $(4 - \sqrt{5})(\sqrt{45} + 2)$ **d** $(5 + \sqrt{2})(6 - \sqrt{8})$

Step into AS

6 Simplify

a $\sqrt{72} + \sqrt{162}$

b $\sqrt{45} - 2\sqrt{5}$

c $\sqrt{50} - \sqrt{8}$

d $\sqrt{75} - \sqrt{48}$

e $2\sqrt{28} + \sqrt{28}$

f $2\sqrt{12} - \sqrt{12} + \sqrt{27}$

Needs more practice ☐ Almost there ☐ I'm proficient! ☐

3.2 Rationalising the denominator

By the end of this section you will know how to:

* Rationalise the denominator of a fraction when the denominator is a surd

AS LINKS

C1: 1.8 Rationalising the denominator of a fraction where it is a surd

Key points

* To **rationalise** the denominator means to remove the surd from the denominator of a fraction.
* To rationalise $\frac{a}{\sqrt{b}}$ you multiply the numerator and denominator by the surd $\sqrt{b}$
* To rationalise $\frac{a}{b + \sqrt{c}}$ you multiply the numerator and denominator by $b - \sqrt{c}$

Guided

1 Rationalise and simplify, if possible.

a $\frac{1}{\sqrt{3}}$

$= \frac{1}{\sqrt{3}} \times \frac{\sqrt{3}}{\sqrt{3}}$

$= \frac{\ldots\ldots}{\ldots\ldots}$

Hint

You must multiply both the numerator and denominator by the same number.

b $\frac{\sqrt{2}}{\sqrt{12}}$

$= \frac{\sqrt{2}}{\sqrt{12}} \times \frac{\sqrt{\ldots\ldots}}{\sqrt{\ldots\ldots}}$

$= \frac{\sqrt{2} \times \ldots\ldots\ldots\ldots}{\ldots\ldots\ldots\ldots}$

$= \frac{\ldots\ldots}{\ldots\ldots}$

c $\frac{3}{2 + \sqrt{5}}$

$= \frac{3}{2 + \sqrt{5}} \times \frac{2 - \sqrt{5}}{\ldots\ldots - \sqrt{\ldots\ldots}}$

$= \frac{3(2 - \sqrt{5})}{\ldots\ldots\ldots\ldots\ldots\ldots}$

$= \frac{3(2 - \sqrt{5})}{\ldots\ldots}$

$= \ldots\ldots\ldots\ldots\ldots\ldots$

$= \ldots\ldots\ldots\ldots\ldots\ldots$

Practice

2 Rationalise and simplify, if possible.

a $\frac{1}{\sqrt{5}}$ **b** $\frac{1}{\sqrt{11}}$ **c** $\frac{2}{\sqrt{7}}$ **d** $\frac{2}{\sqrt{8}}$

e $\frac{2}{\sqrt{2}}$ **f** $\frac{5}{\sqrt{5}}$ **g** $\frac{\sqrt{8}}{\sqrt{24}}$ **h** $\frac{\sqrt{5}}{\sqrt{45}}$

Step into AS

3 Rationalise and simplify.

a $\frac{1}{3-\sqrt{5}}$ **b** $\frac{2}{4+\sqrt{3}}$ **c** $\frac{6}{5-\sqrt{2}}$

Don't forget!

* A surd is ....
* Examples of surds are ....
* $\sqrt{ab} =$ ....
* $\sqrt{\frac{a}{b}} =$ ....
* To rationalise the denominator means to remove the surd from the .... of a fraction.
* To rationalise $\frac{a}{\sqrt{b}}$ you multiply the numerator and denominator by the surd ....
* To rationalise $\frac{a}{b+\sqrt{c}}$ you multiply the numerator and denominator by ....

Exam-style questions

1 Simplify $2\sqrt{45} - \sqrt{80}$

2 Expand and simplify $(3 + \sqrt{2})(5 - \sqrt{18})$

3 Rationalise and simplify $\dfrac{5}{2 - \sqrt{3}}$

4 Rationalise and simplify $\dfrac{3}{\sqrt{5}}$

5 Simplify $\sqrt{72} + \sqrt{50} - \sqrt{32}$

Needs more practice ☐ Almost there ☐ I'm proficient! ☐

4.1 Solving by factorisation

AS LINKS
C1: 2.2 Solving quadratic equations by factorisation

By the end of this section you will know how to:

* Solve quadratic equations by factorising

Key points

* A quadratic equation is an equation in the form $ax^2 + bx + c = 0$ where $a \neq 0$.
* To factorise a quadratic equation find two numbers whose sum is b and whose product is ac.
* When the product of two numbers is 0, then at least one of the numbers must be 0.
* All quadratic equations have two solutions (these may be equal).

Guided

1 Solve

a $5^2 = 15$

$5x^2 - ....= 0$

$5x(.......- ....) = ...$

So $.......= 0$ or $..........= 0$

therefore $x = ...$ or $x = ...$

Hint
Get all terms onto one side of the equation. Do not divide both sides by x. This would lose the solution $x = 0$.

b $x^2 + 7x + 12 = 0$

$(x ....)(x ....) = 0$

So $..........= 0$ or $..........= 0$

therefore $x = ...$ or $x = ...$

c $9x^2 - 16 = 0$

$(3x ....)(.............) = 0$

So $............= 0$ or $............= 0$

therefore $x = ...$ or $x = ...$

d $2x^2 - 5x - 12 = 0$

$(2x ....)(x - 4) = 0$

So $(2x ....) = 0$ or $..........= 0$

therefore $x = ...$ or $x = ...$

Practice

2 Solve

a $6x^2 + 4x = 0$

$x =$ .... or $x =$ ....

b $28x^2 = 21x$

$x =$ .... or $x =$ ....

c $x^2 + 7x + 10 = 0$

$x =$ .... or $x =$ ....

d $x^2 - 5x + 6 = 0$

$x =$ .... or $x =$ ....

e $x^2 - 3x - 4 = 0$

$x =$ .... or $x =$ ....

f $x^2 + 3x - 10 = 0$

$x =$ .... or $x =$ ....

g $x^2 - 10x + 24 = 0$

$x =$ .... or $x =$ ....

h $x^2 - 36 = 0$

$x =$ .... or $x =$ ....

i $x^2 + 3x - 28 = 0$

$x =$ .... or $x =$ ....

j $x^2 - 6x + 9 = 0$

$x =$ .... or $x =$ ....

k $2x^2 - 7x - 4 = 0$

$x =$ .... or $x =$ ....

l $3x^2 - 13x - 10 = 0$

$x =$ .... or $x =$ ....

3 Solve

a $x^2 - 3x = 10$

$x =$ .... or $x =$ ....

b $x^2 - 3 = 2x$

$x =$ .... or $x =$ ....

c $x^2 + 5x = 24$

$x =$ .... or $x =$ ....

d $x^2 - 42 = x$

$x =$ .... or $x =$ ....

e $x(x + 2) = 2x + 25$

$x =$ .... or $x =$ ....

f $x^2 - 30 = 3x - 2$

$x =$ .... or $x =$ ....

g $x(3x + 1) = x^2 + 15$

$x =$ .... or $x =$ ....

h $3x(x - 1) = 2(x + 1)$

$x =$ .... or $x =$ ....

Needs more practice ☐ Almost there ☐ I'm proficient! ☐

4.2 Solving by completing the square

AS LINKS
C1: 2.4 Solving quadratic equations by completing the square

By the end of this section you will know how to:

* Solve quadratic equations by completing the square

Key points

* Completing the square lets you write a quadratic equation in the form $p(x + q)^2 + r = 0$.

Guided

1 Solve $x^2 + 6x + 4 = 0$
Give your solutions in surd form.

$x^2 + 6x + 4 = 0$

$(x + 3)^2 + 4 - \dots\dots = 0$

$(x \dots\dots)^2 - \dots\dots = 0$

$(x \dots\dots)^2 = \dots\dots$

$\dots\dots = \pm\sqrt{\dots\dots}$

$\dots\dots = \dots\dots \pm\sqrt{\dots\dots}$

therefore $x = \dots\dots$ or $x = \dots\dots$

2 Solve $2x^2 - 7x + 4 = 0$
Give your solutions in surd form.

$2\left[x^2 - \frac{7}{2}x + \dots\dots\right] = 0$

$2\left[\left(x - \frac{7}{4}\right)^2 + 2 - \dots\dots\right] = 0$

$(\dots\dots)^2 - \dots\dots = 0$

$(\dots\dots)^2 = \dots\dots$

$\dots\dots = \pm \dots\dots$

$\dots\dots = \pm \dots\dots$

therefore $x = \dots\dots$ or $x = \dots\dots$

Practice

3 Solve by completing the square.

a $x^2 - 4x - 3 = 0$

$x = \dots\dots$ or $x = \dots\dots$

b $x^2 - 10x + 4 = 0$

$x = \dots\dots$ or $x = \dots\dots$

c $x^2 + 8x - 5 = 0$

$x = \dots\dots$ or $x = \dots\dots$

d $x^2 - 2x - 6 = 0$

$x = \dots\dots$ or $x = \dots\dots$

e $2x^2 + 8x - 5 = 0$

$x = \dots\dots$ or $x = \dots\dots$

f $5x^2 + 3x - 4 = 0$

$x = \dots\dots$ or $x = \dots\dots$

Step into AS

4 Solve by completing the square.

a $(x - 4)(x + 2) = 5$ **b** $2x^2 + 6x - 7 = 0$ **c** $x^2 - 5x + 3 = 0$

$x =$ .... or $x =$ .... $x =$ .... or $x =$ .... $x =$ .... or $x =$ ....

Needs more practice ☐ Almost there ☐ I'm proficient! 

4.3 Solving by using the formula

AS LINKS
C1: 2.5 Solving quadratic equations by using the formula

By the end of this section you will know how to:

* Solve quadratic equations by using the formula

Key points

* Any quadratic equation of the form $ax^2 + bx + c = 0$ can be solved using the formula
$$x = \frac{-b \pm \sqrt{b^2 - 4ac}}{2a}$$
* If $b^2 - 4ac$ is negative then the quadratic equation does not have any real solutions.
* It is useful to write down the formula before substituting the values for a, b and c.

Guided

1 Solve $x^2 + 6x + 4 = 0$
Give your solutions in surd form.

Hint
Identify a, b and c and substitute them into the formula.

$a = 1, b = 6, c = 4$

$x = \frac{-(\ldots\ldots) \pm \sqrt{(\ldots\ldots)^2 - 4 \times \ldots\ldots \times \ldots\ldots}}{2 \times \ldots\ldots}$

$x = \frac{\ldots\ldots \pm \sqrt{\ldots\ldots - \ldots\ldots}}{\ldots\ldots}$

$x = \frac{\ldots\ldots \pm \sqrt{\ldots\ldots}}{\ldots\ldots}$

$x = \frac{\ldots\ldots \pm \sqrt{\ldots\ldots \times \ldots\ldots}}{\ldots\ldots}$

$x = \frac{\ldots\ldots + \ldots\ldots\sqrt{\ldots\ldots}}{\ldots\ldots}$ or $x = \frac{\ldots\ldots - \ldots\ldots\sqrt{\ldots\ldots}}{\ldots\ldots}$

The solutions are $x =$ .... or $x =$ ....

2 Solve $3x^2 - 7x - 2 = 0$
Give your solutions in surd form.

$a =$ ...., $b = -7$, $c =$ ....

$x = \frac{\ldots\ldots \pm \sqrt{\ldots\ldots - \ldots\ldots}}{2 \times \ldots\ldots}$

$x = \frac{\ldots\ldots \pm \sqrt{\ldots\ldots + \ldots\ldots}}{\ldots\ldots}$

$x = \frac{\ldots\ldots \pm \sqrt{\ldots\ldots}}{\ldots\ldots}$

$x = \frac{\ldots\ldots + \sqrt{\ldots\ldots}}{\ldots\ldots}$ or $x = \frac{\ldots\ldots - \sqrt{\ldots\ldots}}{\ldots\ldots}$

The solutions are

$x =$ .... or $x =$ ....

Practice

3 Solve, giving your solutions in surd form.

a $3x^2 + 6x + 2 = 0$

$x =$ .... or $x =$ ....

b $2x^2 - 4x - 7 = 0$

$x =$ .... or $x =$ ....

Step into AS

4 Solve, giving your solutions in surd form.

a $4x(x - 1) = 3x - 2$

$x =$ .... or $x =$ ....

b $10 = (x + 1)^2$

$x =$ .... or $x =$ ....

Don't forget!

- To factorise the quadratic equation $ax^2 + bx + c = 0$, find .... numbers whose sum is .... and whose product is ....
- The formula for solving a quadratic equation is $x = \frac{-b \pm \text{.....................}}{\text{.........}}$
- If $b^2 - 4ac$ is .... then the quadratic equation does not have any real roots.

Exam-style questions

1 Solve the equation $x^2 - 7x + 2 = 0$

Give your solutions in the form $\frac{a \pm \sqrt{b}}{c}$, where a, b and c are integers.

2 Solve the equation

$$3x^2 - x - 10 = 0$$

3 Solve $10x^2 + 3x + 3 = 5$

Give your solution in surd form.

5.1 The role of the discriminant

AS LINKS
C1: 2.6 Sketching graphs of quadratic equations

By the end of this section you will know how to:

* Use the discriminant in quadratic equations

Key points

* The formula for solving a quadratic equation $ax^2 + bx + c = 0$ where $a \neq 0$, is $x = \frac{-b \pm \sqrt{b^2 - 4ac}}{2a}$
  The part of the formula $b^2 - 4ac$ is called the **discriminant**.
* If $b^2 - 4ac = 0$, the equation has two real and equal roots (solutions).
* If $b^2 - 4ac > 0$, the equation has two real and distinct roots.
* If $b^2 - 4ac < 0$, the equation has no real roots.

Guided

1 Work out whether the equation $3x^2 + 7x + 5 = 0$ has real and equal, real and distinct or no real roots.

$a =$ ...., $b =$ ...., $c =$ ....

$b^2 - 4ac =$ ....$^2 - 4 \times$ .... $\times$ .... $=$ ....

$=$ ....

Therefore, the equation has .... roots

2 Find the value of p for which $x^2 + 4x + p = 0$ has real and equal roots.

For real and equal roots, $b^2 - 4ac$ 0

$a =$ ...., $b =$ ...., $c =$ ....

$b^2 - 4ac =$ ....$^2 - 4 \times$ .... $\times$ ....

Therefore, $16 -$ .... $= 0$

$4p =$ ....

$p =$ ....

Hint
Form and solve the equation to calculate p.

3 Find the value of h for which $hx^2 + 3x - 7 = 0$ has no real roots.

For no real roots, $b^2 - 4ac$ 0

$a =$ ...., $b =$ ...., $c =$ ....

$b^2 - 4ac =$ ....$^2 - 4 \times$ .... $\times$ .... $=$ ....

Therefore, ... $+$ .... < 0

.... $<$ ....

$h <$ ....

Practice

Work out whether each of the equations has two real and equal, two real and distinct or no real roots.

4 $4x^2 - 5x + 7 = 0$

5 $6x^2 - 2x - 3 = 0$

6 $9x^2 - 30x + 25 = 0$

7 $2x^2 + 2x + 7 = 0$

8 Find the values of q for which $x^2 + qx + 16 = 0$ has real and equal roots.

9 Find the values of q for which $2x^2 + qx + 9 = 0$ has real and equal roots.

10 Find the values of r for which $rx^2 - 10x + r = 0$ has real and equal roots.

11 Find the values of t for which $4x + t = 3x^2$ has real and distinct roots.

Needs more practice ☐ Almost there ☐ I'm proficient! ☐

5.2 The sum and product of the roots of a quadratic equation

By the end of this section you will know how to:

* Find the sum and product of the roots of $ax^2 + bx + c = 0$ from the values of its coefficients a, b and c
* Use the sum and product of the roots to write the corresponding quadratic equation

Key points

* The **sum** of the roots of the quadratic equation $ax^2 + bx + c = 0$, $a \neq 0$, is given by $-\frac{b}{a}$
* The **product** of the roots of quadratic equation $ax^2 + bx + c = 0$, $a \neq 0$, is given by $\frac{c}{a}$
* The quadratic equation can be written as:
  $x^2 -$ (sum of the roots)x + (product of the roots) = 0 or $x^2 - \left(-\frac{b}{a}\right)x + \left(\frac{c}{a}\right) = 0$

Guided

1 Find the sum and product of the roots of the equation $2x^2 + 6x - 5 = 0$.

$a =$ ...., $b =$ ...., $c =$ ....

Sum $= -\frac{b}{a} = \frac{.......}{.......} =$ ....

Product $= \frac{c}{a} = \frac{.......}{.......} =$ ....

2 The sum of the roots of a quadratic equation is −7 and the product of its roots is 10.
Write down the quadratic equation.

$-\frac{b}{a} =$ ...., $\frac{c}{a} =$ ....

Substituting these values into $x^2 - \left(-\frac{b}{a}\right)x + \left(\frac{c}{a}\right) = 0$ gives:

$x^2 - ($.......$)x + ($.......$) = 0$ or ....

Practice

Find the sum and product of the roots of the equations.

3 $x^2 - 11x + 30 = 0$

4 $5x^2 + 8x - 21 = 0$

5 $9x^2 - 16 = 0$

6 $6x^2 + x - 15 = 0$

Write the quadratic equation when the sum and product of its roots are

7 -2 and -8

8 $-\frac{1}{3}$ and $-\frac{2}{3}$

9 -8.5 and -4.5

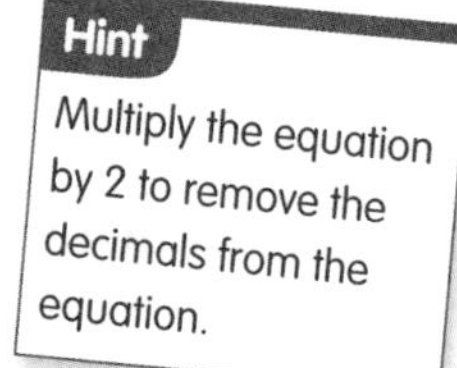

10 -1.5 and -14.5

Don't forget!

- The formula for solving a quadratic equation $ax^2 + bx + c = 0$ where $a \neq 0$, is

  .... = $\frac{\text{..}}{\text{.............}}$

- The part of the formula $b^2 - 4ac$ is called the ....
- If $b^2 - 4ac = 0$, the equation has ....
- If $b^2 - 4ac > 0$, the equation has ....
- If $b^2 - 4ac < 0$, the equation has ....
- The sum of the roots of the quadratic equation $ax^2 + bx + c = 0$, $a \neq 0$, is given by ....
- The product of the roots of quadratic equation $ax^2 + bx + c = 0$, $a \neq 0$, is given by ....
- The quadratic equation can be written as ....

Exam-style questions

1 Find the values of g for which $3x^2 + gx + 16 = 0$ has equal roots.

2 For a quadratic equation

the sum of its roots is -2.5

the product of its roots is 4.5

Find the quadratic equation in the form $ax^2 + bx + c = 0$

where a, b and c are integers.

Needs more practice ☐ Almost there ☐ I'm proficient! ☐

6.1 Solving simultaneous linear equations using elimination

AS LINKS
C1: 3.1 Solving simultaneous linear equations by elimination

By the end of this section you will know how to:

* Solve two simultaneous linear equations using the elimination method

Key points

* Two equations are simultaneous when they are both true at the same time.
* Solving simultaneous linear equations in two unknowns involves finding the value of each unknown which works for both equations.
* Make sure that the coefficient of one of the unknowns is the same in both equations.
* Eliminate this equal unknown by either subtracting or adding the two equations.

Guided

Solve these simultaneous equations.

1 $3x + y = 5$
$x + y = 1$

Subtracting the second equation from the first equation eliminates the y term to give:

$2x =$ ....

$x =$ ....

$y =$ ....

Hint
To find the second unknown, substitute your value for the first unknown into one of the original equations. Then check your solutions by substituting the values for x and y into the other equation.

2 $x + 2y = 13$
$5x - 2y = 5$

Adding both of these equations together eliminates the y term to give:

$6x =$ ....

$x =$ ....

$y =$ ....

Practice

Solve these simultaneous equations.

3 $4x + y = 8$
$x + y = 5$

4 $3x + y = 7$
$3x + 2y = 5$

5 $4x + y = 3$
$3x - y = 11$

6 $3x + 4y = 7$
$x - 4y = 5$

Step into AS

Solve these simultaneous equations.

7 $2x + y = 11$
$x - 3y = 9$

8 $2x + 3y = 11$
$3x + 2y = 4$

Needs more practice ☐ Almost there ☐ I'm proficient! ☐

6.2 Solving simultaneous linear equations using substitution

AS LINKS
C1: 3.2 Solving simultaneous linear equations by substitution

By the end of this section you will know how to:

- Solve two simultaneous linear equations using the substitution method

Key points

- To solve simultaneous linear equations in two unknowns involves finding the value of each unknown which works for both equations.
- The substitution method used here will help in section 6.3.

Guided

Solve the simultaneous equations.

1 $y = 2x + 1$ (equation 1)
$5x + 3y = 14$ (equation 2)

Substituting $2x + 1$ for y into equation 2 gives

$5x + 3(\dots\dots) = 14$

$5x + \dots\dots = 14$

$11x = \dots\dots$

$x = \dots\dots$

$y = \dots\dots$

Hint
To find the second unknown, substitute your value for the first unknown into one of the original equations.

2 $2x - y = 16$ (equation 1)
$4x + 3y = -3$ (equation 2)

Rearranging equation 1 gives

$y = 2x - 16$

Substituting $2x - 16$ for y into equation 2 gives

$4x + 3(\dots\dots) = -3$

$4x + \dots\dots = -3$

$\dots\dots = \dots\dots$

$x = \dots\dots$

$y = \dots\dots$

Practice

Solve these simultaneous equations.

3 $y = x - 4$
$2x + 5y = 43$

4 $y = 2x - 3$
$5x - 3y = 11$

5 $2y = 4x + 5$
$9x + 5y = 22$

6 $2x = y - 2$
$8x - 5y = -11$

Solve these simultaneous equations.

7 $3x + 4y = 8$
$2x - y = -13$

8 $3y = 4x - 7$
$2y = 3x - 4$

9 $3x = y - 1$
$2y - 2x = 3$

10 $3x + 2y + 1 = 0$
$4y = 8 - x$

Needs more practice ☐ Almost there ☐ I'm proficient! ☐

Solving simultaneous equations where one is quadratic

AS LINKS

C1: 3.3 Using substitution when one equation is linear and the other is quadratic

By the end of this section you will know how to:

* Solve simultaneous linear and quadratic equations

Key points

* Make one of the unknowns the subject of the linear equation (rearranging where necessary).
* Use this to substitute into the quadratic equation.
* There are usually two pairs of solutions.

Guided

Solve these simultaneous equations.

1 $y = x + 1$
$x^2 + y^2 = 13$

$x^2 + (\text{..........})^2 = 13$

$x^2 + \text{....................} - 13 = 0$

$2x^2 + \text{..............} = 0$

$\text{....................} = 0$

$(\text{..........})(\text{..........}) = 0$

$x = \text{......}$ or $x = \text{......}$

When $x = \text{......}$, $y = \text{......}$

When $x = \text{......}$, $y = \text{......}$

Hint
Substitute the linear equation into the quadratic (you may need to rearrange first).

2 $2x + 3y = 5$
$2y^2 + xy = 12$

$x = \text{..............}$

$2y^2 + y(\text{..............}) = 12$

$2y^2 + \text{..............} - 12 = 0$

$\text{....................................} = 0$

$\text{....................} = 0$

$(\text{..........})(\text{..........}) = 0$

$y = \text{......}$ or $y = \text{......}$

When $y = \text{......}$, $x = \text{......}$

When $y = \text{......}$, $x = \text{......}$

Practice

Solve these simultaneous equations.

3 $y = x + 5$
$x^2 + y^2 = 25$

4 $y = 2x - 1$
$x^2 + xy = 24$

5 $y = 2x$
$y^2 - xy = 8$

6 $2x + y = 11$
$xy = 15$

Solve the simultaneous equations, giving your answer in their simplest surd form.

7 $x - y = 1$
$x^2 + y^2 = 3$

8 $y - x = 2$
$x^2 + xy = 3$

Don't forget!

* Simultaneous linear equations can be solved either by the .... method or by the .... method.
* There are usually .... pairs of solutions when you solve simultaneous linear and quadratic equations.

Exam-style questions

1 Solve these simultaneous equations.

$2x + 3y = 2$
$5x + 4y = 12$

2 Solve these simultaneous equations.

$3y + x = 4$
$x^2 - y^2 = 6$

Needs more practice ☐ Almost there ☐ I'm proficient! ☐

7.1 General (nth) term of arithmetic series

AS LINKS
C1: 6.5 Arithmetic series

By the end of this section you will know how to:

* Find the nth term of an arithmetic series
* Use the nth term of an arithmetic series

Key points

* A **series** is formed when the terms of a sequence are added together.
* The **general term** of a series (or sequence) is commonly called the **nth term**.
* An arithmetic series (or sequence) is one where each term in the series (or sequence) increases by the same amount.
* The nth term (general term) of an arithmetic series is $a + (n - 1)d$, where a is the first term and d is the **common difference** (the amount each term increases by).

Guided

1 Find the first five terms of the series for which the nth term is $4n + 1$.

First term = $4 \times 1 + 1 =$ ....

Second term = $4 \times$ .... + ... = ...

Third term = ... $\times$ .... + ... = ...

Fourth term = ... $\times$ .... + ... = ...

Fifth term = ... $\times$ .... + ... = ...

So the first five terms of the series $4n + 1$ are 5, 9, ..., ..., ...

Hint
'First term' means that $n = 1$, 'second term' means that $n = 2$, etc.

2 The nth term of an arithmetic series is $5n - 2$. Which term has a value of 73?

$5n - 2 = 73$

$5n =$ ....

$n =$ ....

3 Find the nth term of the arithmetic series $3 + 8 + 13 + 18 + \ldots$

nth term $= a + (n - 1)d$

$a =$ ...., $d =$ ....

nth term = ... $+ (n - 1) \times$ ....

$= 3 +$ .... $- 5$

$=$ ....

Practice

4 Find the first three terms of the arithmetic series when the nth term is $5n + 3$.

5 Find the nth and 20th terms of the arithmetic series $5 + 8 + 11 + 14 + \ldots$

6 Find the nth and 10th terms of the arithmetic series $15 + 13 + 11 + 9 + \ldots$

7 Find the 20th and 100th terms of the arithmetic series $6 + 10 + 14 + 18 + \ldots$

8 Find the 15th and 50th terms of the arithmetic series $50 + 47 + 44 + 41 + \ldots$

9 Find n, the number of terms in the arithmetic series $5 + 8 + 11 + 14 + \ldots + 77$

10 Find n, the number of terms in the arithmetic series $70 + 62 + 54 + 46 + \ldots + (-346)$

11 The first term of an arithmetic series is 2. The fifth term is 22.
What is the common difference?

12 The fourth term of an arithmetic series is 10. The seventh term is 19.
Find the first term and the common difference.

Needs more practice ☐ Almost there ☐ I'm proficient! ☐

7.2 The sum of an arithmetic series

AS LINKS
C1: 6.6 The sum to n terms of an arithmetic series

By the end of this section you will know how to:

- Find the sum of an arithmetic series
- Use the sum of an arithmetic series

Key points

- You find the sum, S_n, of an arithmetic series using the formula $S_n = \frac{n}{2}[2a + (n - 1)d]$ where a is the first term, d is the common difference and n is the number of terms.
- Alternatively, you can use the formula $S_n = \frac{n}{2}(a + L)$ where a is the first term, n is the number of terms and L is the last term.

Guided

1 Find the sum of the arithmetic series 1 + 5 + 9 + 13 + … with 30 terms.

$S_n = \frac{n}{2}[2a + (n - 1)d]$

$a = 1, d = 4, n =$ ....

$S_n = \frac{\dots}{2}[2 \times$ .... $+ ($ .... $- 1) \times$ .... $]$

$S_n = 15 \times ($ .... $+$ .... $\times$ .... $)$

$S_n =$ ....

2 An arithmetic series has a first term of 7 and a last term of 41.
Work out the number of terms which give a sum of 432.

$S_n = \frac{n}{2}(a + L)$

$S_n =$ ...., $a =$ ...., $L =$ ....

.... $= \frac{n}{2}($ .... + ... $)$

.... $=$ .... n

$n =$ ....

3 An arithmetic series begins 7 + 9 + 11 + 13 + ...
Work out the number of terms which give a sum of 352.

$S_n = \frac{n}{2}[2a + (n - 1)d]$

$S_n =$ ...., $a =$ ...., $d =$ ....

.... $= \frac{n}{2}[2 \times$ .... $+ ($ .... $- 1) \times$ .... $]$

$704 = n($ .... $+ 2n -$ .... $)$

$704 =$ ....

$2n^2 +$ .... $= 0$

$n^2 + 6n - 352 = 0$

(................)(................) = 0

$n =$ ....

Hint
Form a quadratic equation, factorise and solve it. Remember that you can only have positive values of n.

Hint
You may be able to get a simpler equation by cancelling earlier in your working.

4 The sum of the first two terms of an arithmetic series is 1.
The 20th term is 93. Find the first term and the common difference of the series.

From the information that the sum of the first two terms = 1:

$S_n =$ ...., $n =$ ....

$S_n = \frac{n}{2}[2a + (n - 1)d]$

.... $= \frac{......}{2}[2a + ($ .... $- 1)d]$

$2a +$ .... $= 1$ (equation 1)

Hint
Form linear equations and solve them simultaneously.

From the information that the 20th term = 93:

$n = 20$

nth term $= a + (n - 1)d$

$93 = a + ($ .... $- 1)d$

$a +$ .... $= 93$ (equation 2)

Solving the equations 1 and 2 simultaneously gives $a =$ ...., $d =$ ....

So first term = ..., common difference = ...

5 Find the sum of the first 20 terms of the arithmetic series $2 + 5 + 8 + \ldots$

6 Find the sum of the first 30 terms of the arithmetic series $3 + 6 + 9 + \ldots$

7 Find the sum of the first 50 terms of the arithmetic series $40 + 34 + 28 + \ldots$

8 Find the sum of the first 20 terms of the arithmetic series $100 + 91 + 82 + \ldots$

9 Find the sum of all the odd numbers from 21 to 41 inclusive.

10 Find the sum of the arithmetic series where the first term is 2, the last term is 135 and there are 20 terms.

11 Find the sum of the arithmetic series where the first term is 8, the last term is 53 and there are 16 terms.

12 The sum to n terms of the arithmetic series $3 + 5 + 7 + \ldots$ is 120. Find the value of n.

13 The sum of the first three terms of an arithmetic series is 15.
The 10th term is 29. Find the first term and the common difference of the series.

Don't forget!

* A series is formed when the terms of a .... are added together.
* The general term of a series (or sequence) is commonly called the ....
* An arithmetic series (or sequence) is one where each term in the series (or sequence) increases by ....
* The nth term of an arithmetic series is ...., where a is the first term and d is the common difference.
* You find the sum, S_n, of an arithmetic series using the formula $S_n =$ .... where a is the first term, d is the common difference and n is the number of terms.
* Alternatively, you can use the formula $S_n =$ .... where a is the first term, n is the number of terms and L is the last term.

Exam-style questions

1 The sum of the first four terms of an arithmetic series is 198. The 20th term of this series is -73

a Find the first term of the series and the common difference.

b Find the sum of the first 30 terms of the series.

8.1 The equation of a line

AS LINKS
C1: 5.4 The formula for finding the equation of a straight line

By the end of this section you will know how to:

- Find the gradient and y-intercept of a line from its equation
- Find the gradient and the equation of a straight line

Key points

- A straight line has the equation $y = mx + c$, where m is the gradient and c is the y-intercept (where $x = 0$).
- The equation of a straight line can be written in the form $ax + by + c = 0$, where a, b and c are integers.
- When given the coordinates (x_1, y_1) and (x_2, y_2) of two points on a line the gradient is calculated using the formula $m = \frac{y_2 - y_1}{x_2 - x_1}$

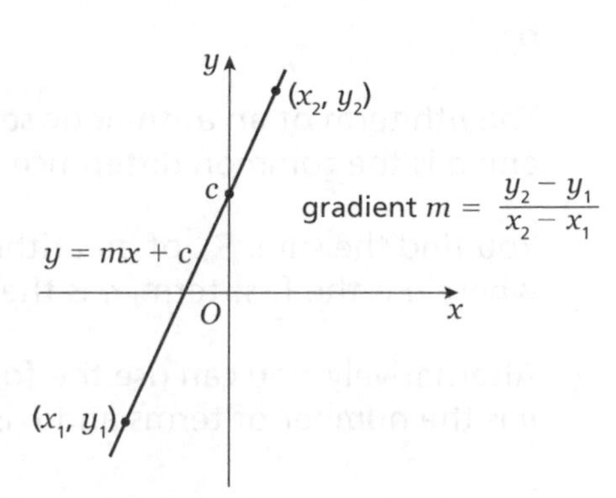

Guided

1 A straight line has gradient $-\frac{1}{2}$ and y-intercept 3.
Write the equation of the line in the form $ax + by + c = 0$.

$m = -\frac{1}{2}$

$c = 3$

$y =$ ....

....

.... $= 0$

Hint
Rearrange the equation in the form $y = mx + c$.

2 Find the gradient and the y-intercept of the line with the equation $3y - 2x + 4 = 0$.

$3y =$ ....

$y =$ ....

gradient $= m =$ ....

y-intercept $= c =$ ....

3 Find the equation of the line which passes through the point (5, 13) and has gradient 3.

$m =$ ....

$y = mx + c$

$y =$ .... $x + c$

Substituting $x = 5$, $y = 13$ into the equation gives

.... = ... × .... $+ c$

.... = ... $+ c$

$c =$ ....

The equation is $y =$ .... $x -$ ....

4 Find the equation of the line passing through the points with coordinates (2, 4) and (8, 7).

$m = \frac{y_2 - y_1}{x_2 - x_1}$

$m = \frac{\ldots\ldots - 4}{\ldots\ldots - 2} = \frac{\ldots\ldots}{\ldots\ldots} = \ldots\ldots$

$y = \ldots\ldots x + c$

Substituting the coordinates of either point into the equation gives

$\ldots\ldots = \ldots\ldots \times \ldots\ldots + c$

$\ldots\ldots = \ldots\ldots + c$

$c = \ldots\ldots$

$y = \ldots\ldots x + \ldots\ldots$

Practice

5 Find the gradient and the y-intercept of the following equations.

a $y = 3x + 5$

b $y = -\frac{1}{2}x - 7$

c $2y = 4x - 3$

d $x + y = 5$

e $2x - 3y - 7 = 0$

f $5x + y - 4 = 0$

6 Complete the table, giving the equation of the line in the form $y = mx + c$.

Gradient	y-intercept	Equation of the line
5	0	
−3	2	
4	−7	

7 Find, in the form $ax + by + c = 0$ where a, b and c are integers, an equation for each of the following lines.

a gradient $-\frac{1}{2}$, y-intercept −7

b gradient 2, y-intercept 0

c gradient $\frac{2}{3}$, y-intercept 4

d gradient −1.2, y-intercept −2

8 Write an equation for the line which passes through the point (2, 5) and has gradient 4.

9 Write an equation for the line which passes through the point (6, 3) and has gradient $-\frac{2}{3}$.

10 Write an equation for the line passing through each of the following pairs of points.

a (4, 5), (10, 17)

b (0, 6), (−4, 8)

c (−1, −7), (5, 23)

d (3, 10), (4, 7)

8.2 Parallel and perpendicular lines

AS LINKS
C1: 5.5 The conditions for two straight lines to be parallel or perpendicular

By the end of this section you will know how to:

* Work out the gradient of a line which is parallel to or perpendicular to a given line.
* Form the equations of lines which are parallel to or perpendicular to a given line.

* When lines are **parallel** they have the same gradient.
* A line **perpendicular** to the line with equation $y = mx + c$ has gradient $-\frac{1}{m}$.

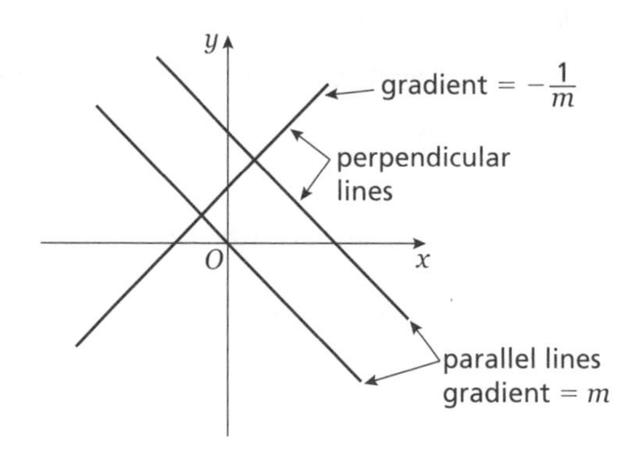

Guided

1 Find the equation of the line parallel to $y = 2x + 4$ which passes through the point (4, 9).

$m =$ ....

$y =$ .... $x + c$

Substituting the coordinates into the equation gives

.... = ... × .... $+ c$

$c =$ ....

$y =$ ....

2 Find the equation of the line perpendicular to $y = 2x - 3$ which passes through the point (−2, 5).

For the line $y = 2x - 3$

$m =$ ....

$-\frac{1}{m} =$ ....

$y =$ .... $x + c$

Substituting the coordinates into the equation gives

.... = ... × .... $+ c =$ ....

$c =$ ....

$y =$ ....

3 Find the equation of the line perpendicular to $y = \frac{1}{2}x - 3$ which passes through the point (−5, 3).

For the line $y = \frac{1}{2}x - 3$

$m =$ ....

$-\frac{1}{m} =$ ....

$y =$ .... $x + c$

Substituting the coordinates into the equation gives

.... = ... × .... $+ c$

$c =$ ....

$y =$ ....

Hint
If $m = \frac{a}{b}$, then the negative reciprocal $-\frac{1}{m} = -\frac{b}{a}$

4 A line passes through the points (0, 5) and (9, −1).
Find the equation of the line which is perpendicular to the line and passes through its midpoint.

$m = \frac{\ldots\ldots - \ldots\ldots}{\ldots\ldots - \ldots\ldots} = \frac{\ldots\ldots}{\ldots\ldots} = \ldots\ldots\ldots\ldots$

$-\frac{1}{m} = \ldots\ldots\ldots\ldots$

$y = \ldots\ldots\ldots\ldots x + c$

The coordinates of the midpoint of the line are

$\left(\frac{0+9}{2}, \frac{5+-1}{2}\right) = \left(\frac{\ldots\ldots}{\ldots\ldots}, \ldots\ldots\right)$

Substituting the coordinate of the midpoint into the equation $y = \ldots\ldots\ldots\ldots x + c$ gives

$\ldots\ldots = \frac{\ldots\ldots}{\ldots\ldots} \times \frac{\ldots\ldots}{\ldots\ldots} + c = \ldots\ldots\ldots\ldots\ldots\ldots$

$c = \ldots\ldots\ldots\ldots$

$y = \ldots\ldots\ldots\ldots\ldots\ldots$

5 Find the equation of the line parallel to each of the given lines and which passes through each of the given points.

a $y = 3x + 1$ (3, 2)

b $y = 3 - 2x$ (1, 3)

c $2x + 4y + 3 = 0$ (6, −3)

d $2y - 3x + 2 = 0$ (8, 20)

6 Find the equation of the line perpendicular to each of the given lines and which passes through each of the given points.

a $y = 2x - 6$ (4, 0)

b $y = -\frac{1}{3}x + \frac{1}{2}$ (2, 13)

c $x - 4y - 4 = 0$ (5, 15)

d $5y + 2x - 5 = 0$ (6, 7)

7 In each case find an equation for the line passing through the origin which is also perpendicular to the line joining the two points given.

a (4, 3), (−2, −9)

b (0, 3), (−10, 8)

8 Work out whether these pairs of lines are parallel, perpendicular or neither.

a $y = 2x + 3$
$y = 2x - 7$

b $y = 3x$
$2x + y - 3 = 0$

c $y = 4x - 3$
$4y + x = 2$

d $3x - y + 5 = 0$
$x + 3y = 1$

e $2x + 5y - 1 = 0$
$y = 2x + 7$

f $2x - y = 6$
$6x - 3y + 3 = 0$

Don't forget!

- A straight line has the equation ...., where m is the gradient and c is the y-intercept.
- The equation of a straight line can be written in the form ...., where a, b and c are integers.
- When given the coordinates (x_1, y_1) and (x_2, y_2) of two points on a line the gradient is calculated using the formula $m = \frac{\text{.....................}}{\text{.....................}}$
- When lines are parallel they have the same ....
- A line perpendicular to $y = mx + c$ has gradient ....

Exam-style questions

1 The straight line $\mathbf{L_1}$ passes through the points A and B with coordinates $(-4, 4)$ and $(2, 1)$, respectively.

a Find an equation of $\mathbf{L_1}$ in the form $ax + by + c = 0$

....

The line $\mathbf{L_2}$ is parallel to the line $\mathbf{L_1}$ and passes through the point C with coordinates $(-8, 3)$.

b Find an equation of $\mathbf{L_2}$ in the form $ax + by + c = 0$

....

The line $\mathbf{L_3}$ is perpendicular to the line $\mathbf{L_1}$ and passes through the origin.

c Find an equation of $\mathbf{L_3}$

....

Needs more practice ☐ Almost there ☐ I'm proficient! ☐

9.1 Recognising graphs

AS LINKS
C2: 8.4 Recognise the graphs of sin θ, cos θ and tan θ

By the end of this section you will know how to:

- Recognise graphs of linear, quadratic, cubic, reciprocal, exponential, and circular functions
- Draw and understand tangents and normals to graphs

Key points

- The graph of the **linear** function $y = mx + c$ is a straight line.
- The graph of the **quadratic** function $y = ax^2 + bx + c$, where $a \neq 0$, is a curve called a **parabola**. Parabolas have a line of symmetry.

- The graph of a **cubic** function, which can be written in the form $y = ax^3 + bx^2 + cx + d$, where $a \neq 0$, has one of these shapes:

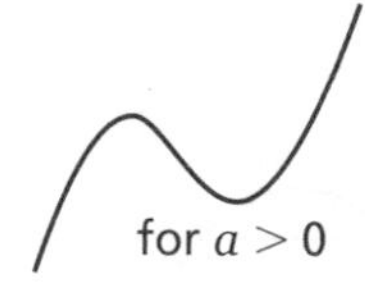

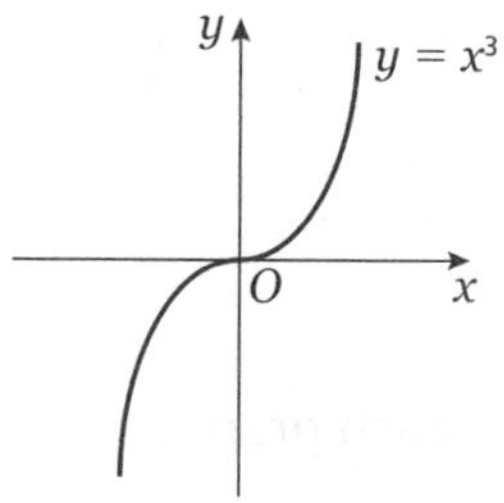

special case: $a = 1$

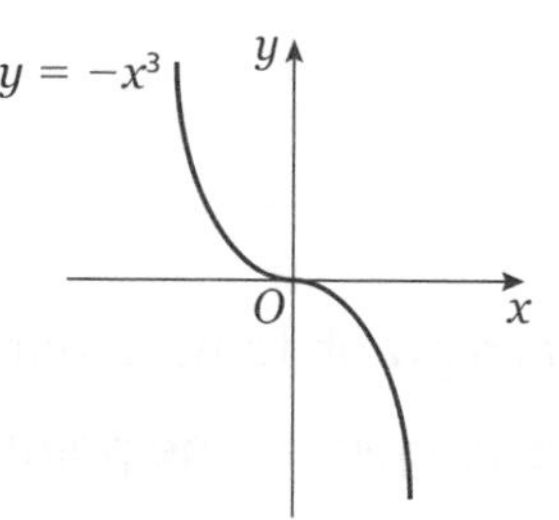

special case: $a = -1$

- The graph of a **reciprocal** function of the form $y = \frac{a}{x}$ has these shapes:

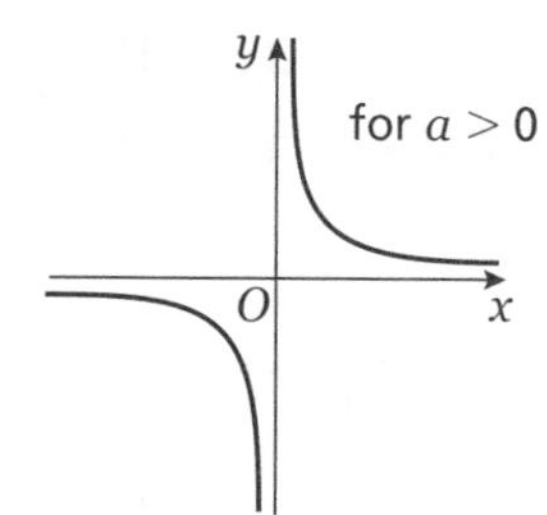

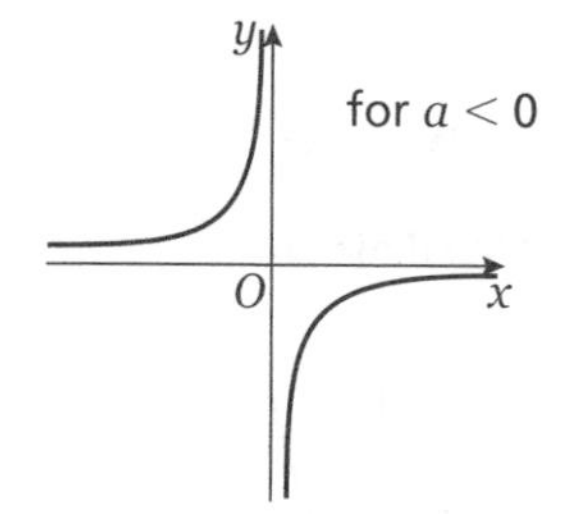

- An **exponential** function is of the form $y = a^x$, where $a > 0$. The graph of an exponential has one of these shapes:

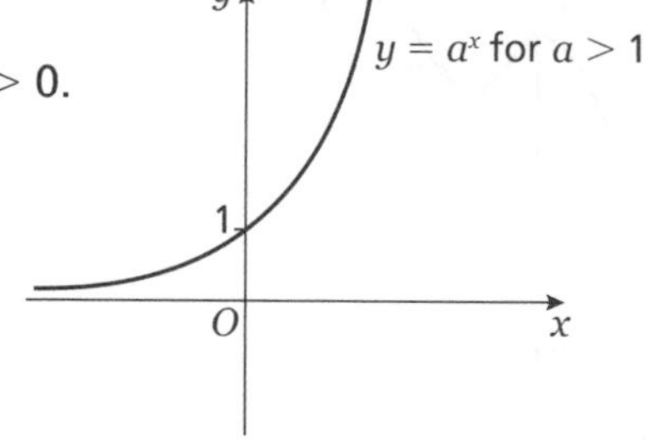

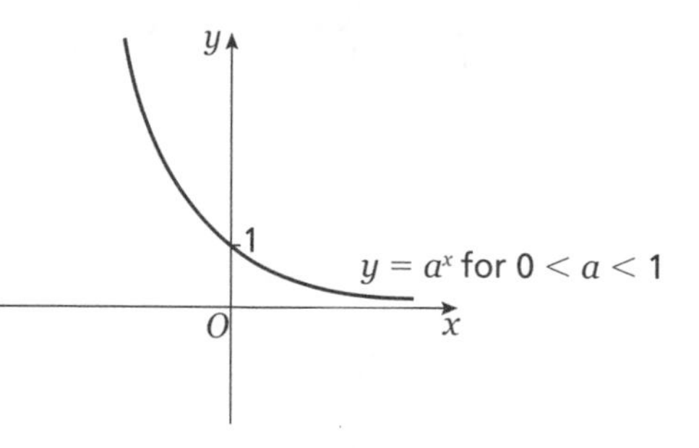

- **Circular** (or trigonometric) functions include sine, cosine and tangent. Their graphs have these shapes:

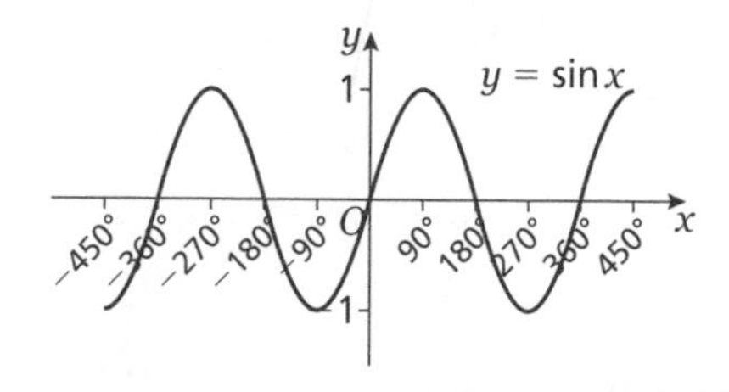

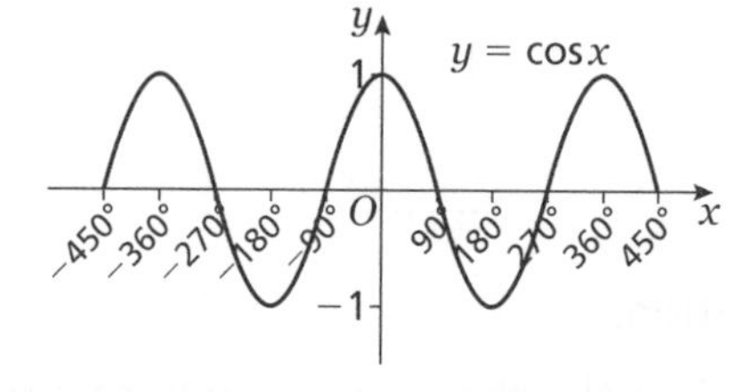

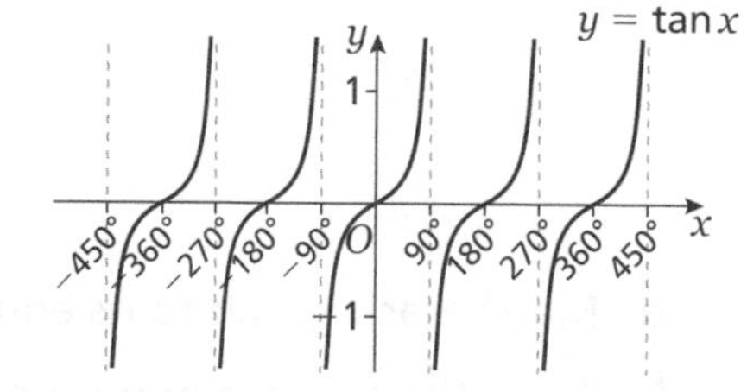

* The **tangent** to a curve is a straight line which touches the curve but does not cross it.
* The **normal** to a curve is perpendicular to the tangent at that point on the curve:

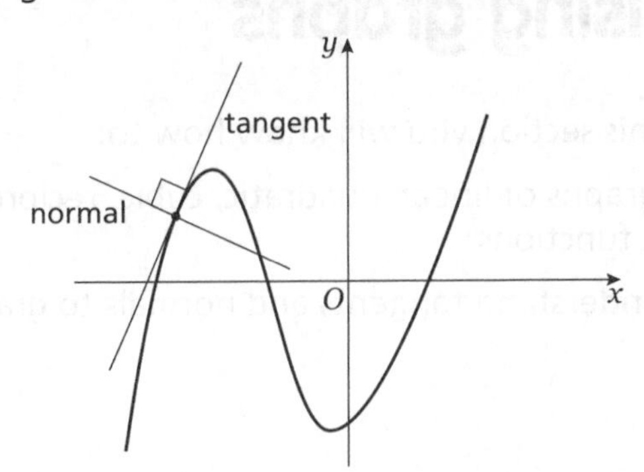

Guided

1 Here are three equations.

A $y = 3x - x^2$ **B** $y = 2^x$ **C** $y = x^2 + 2$

Hint
Two of the equations are quadratics.

Here are three sketch graphs.

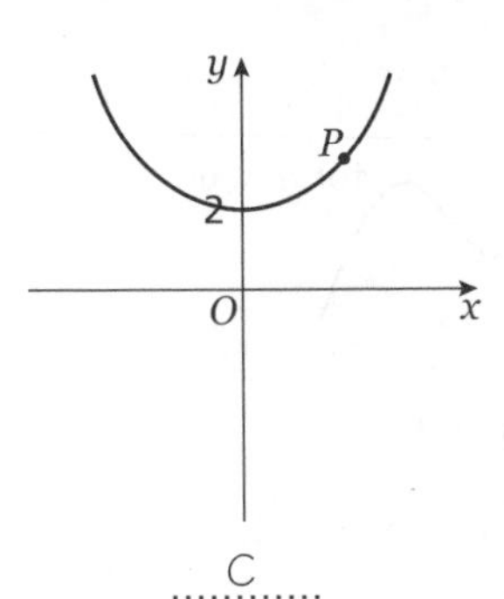

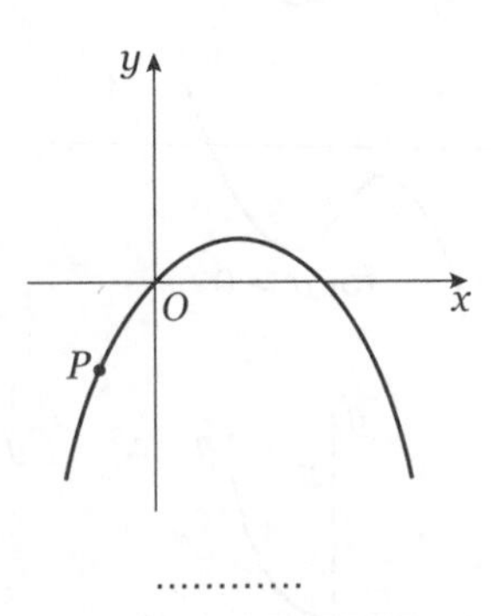

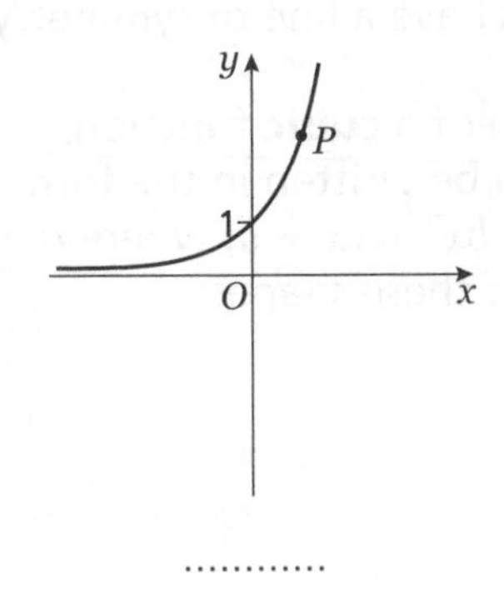

C ....

a Match each graph to its equation.

b Draw the tangent at the point P on each graph.

Practice

2 Here are six equations.

A $y = \frac{5}{x}$ **B** $y = x^2 + 3x - 10$ **C** $y = x^3 + 3x^2$

D $y = 1 - 3x^2 - x^3$ **E** $y = x^3 - 3x^2 - 1$ **F** $x + y = 5$

Hint
Find where each of the cubic equations cross the y-axis.

Here are six graphs.

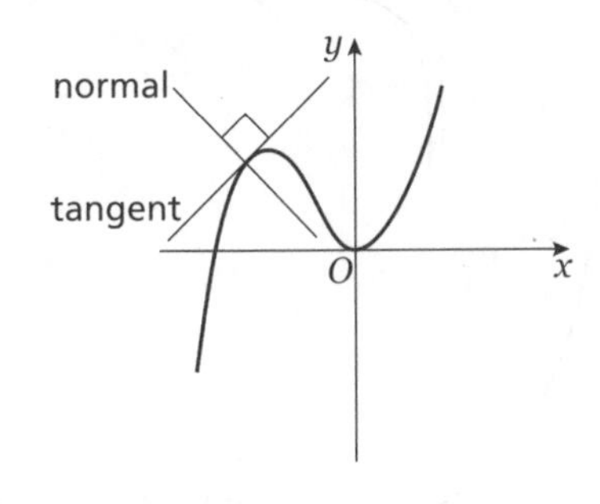

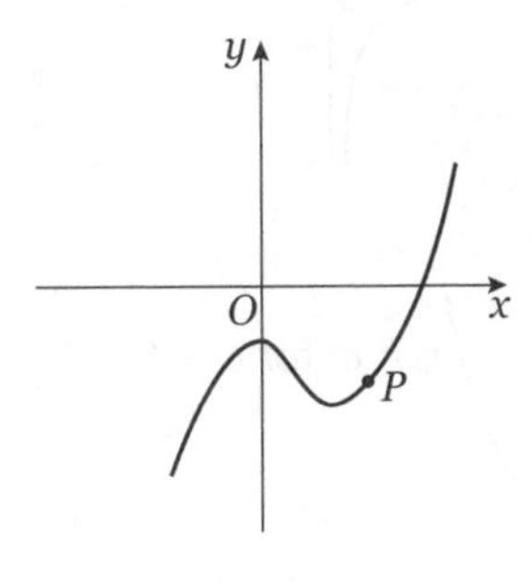

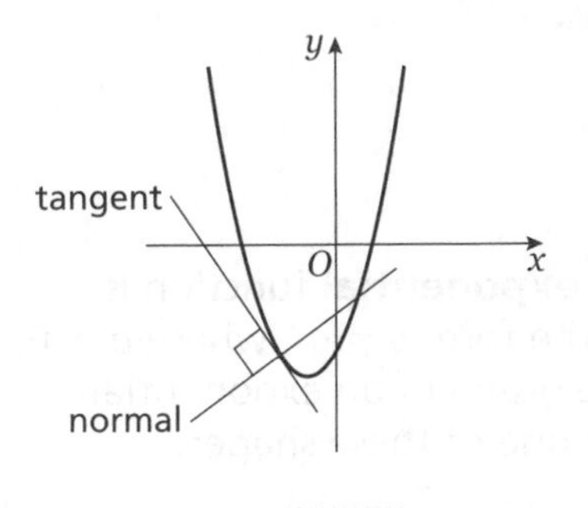

....

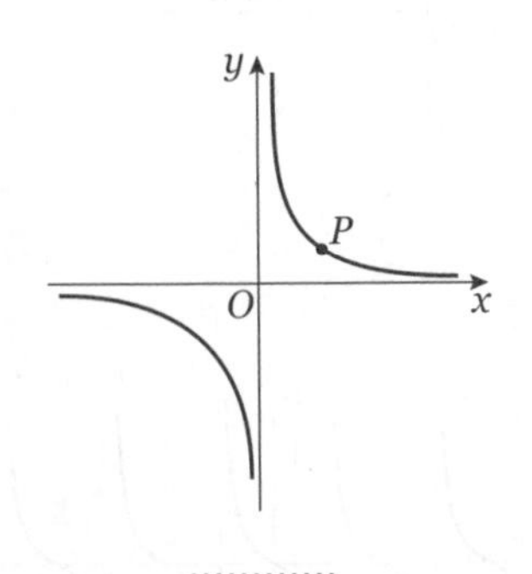

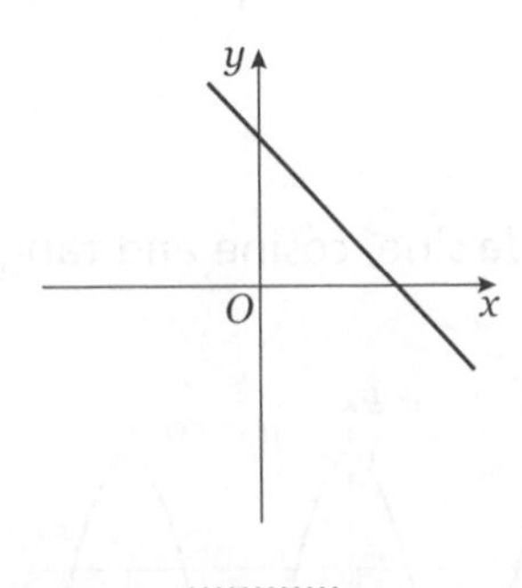

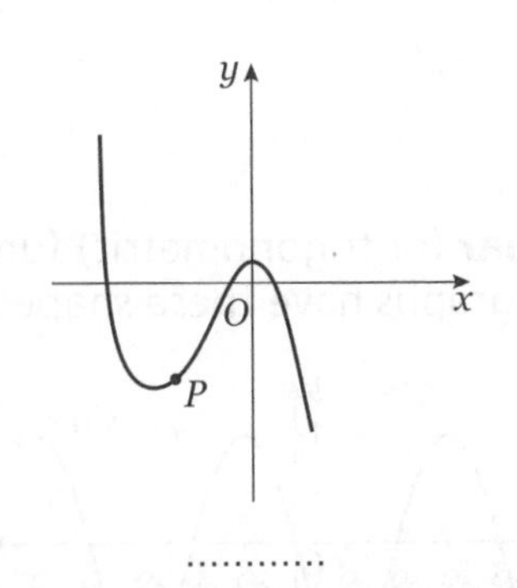

....

a Match each graph to its equation.

b Draw the tangent and normal on the three graphs where a point P is marked.

Step into AS

3 Here are three equations.

A $y = 2\tan x$ **B** $y = 3\sin x$ **C** $y = -\cos x$

Here are three graphs.

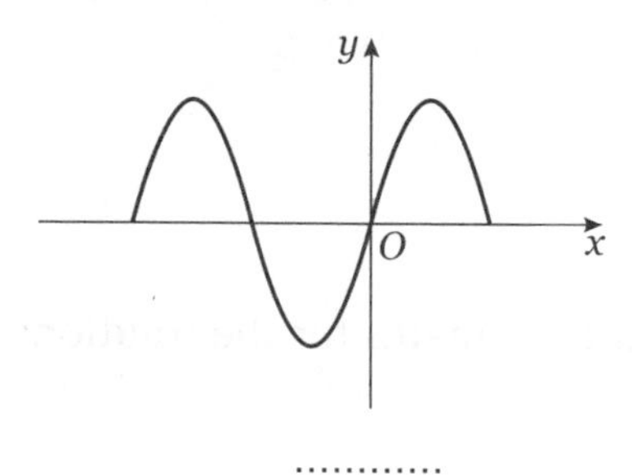

....

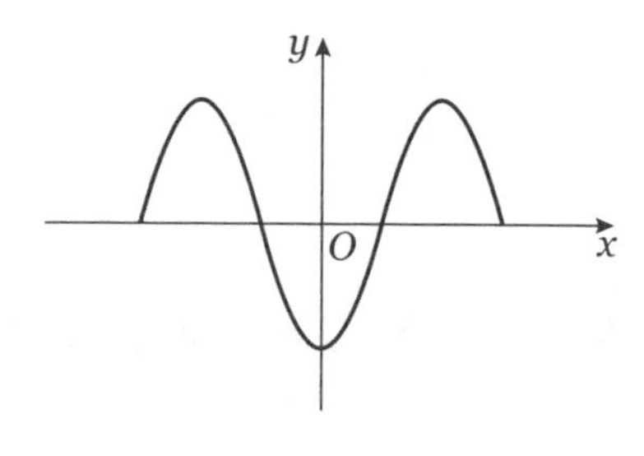

....

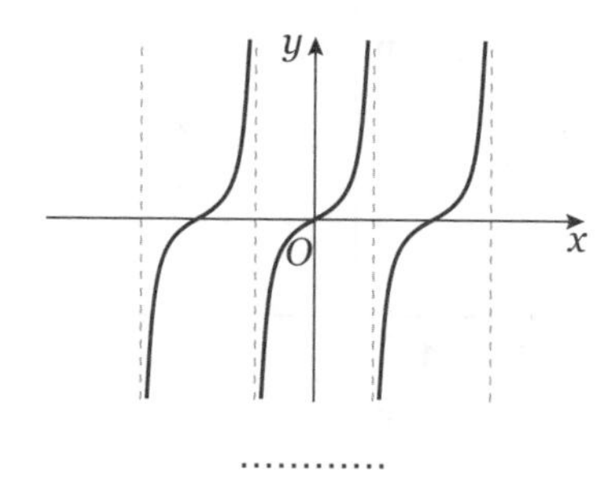

....

Match each graph to its equation.

Needs more practice ☐ Almost there ☐ I'm proficient! ☐

9.2 Drawing and using graphs

AS LINKS
C1: 2.1 Plotting the graphs of quadratic functions

By the end of this section you will know how to:

* Draw graphs of linear, quadratic and cubic functions
* Use the graphs of linear, quadratic and cubic functions to solve equations

Key points

* To draw a linear, quadratic or cubic graph, calculate and plot the coordinates of a number of points and then join them up.

Guided

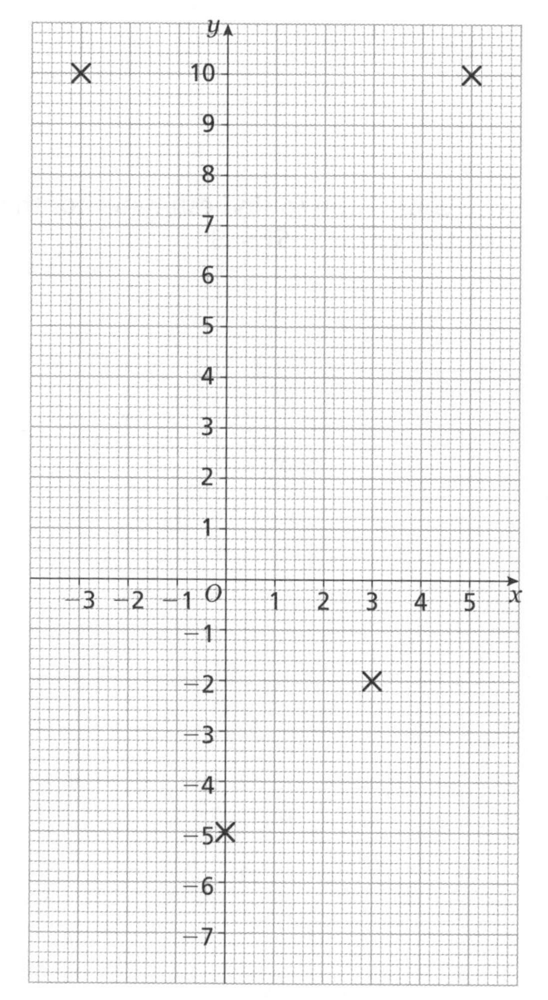

1 a On the grid left, draw the graph of $y = x^2 - 2x - 5$ for values of x from -3 to 5.

x	-3	-2	-1	0	1	2	3	4	5
y	10			-5			-2		10

Hint
Draw and complete a table to find the coordinates of points on the graph.

Hint
Join the points with a smooth curve. Quadratic curves are symmetrical.

b Use your graph to find estimates for the solutions of $x^2 - 2x - 5 = 0$.

Using the graph, the solutions to the equation are when $y = 0$, which is where the curve intersects the x-axis.

This gives the solutions

$x \approx$ .... or $x \approx$ ....

Practice

2 a On the grid below, draw the graph of $y = 3 + 4x - 2x^2$ for values of x from -2 to 4.

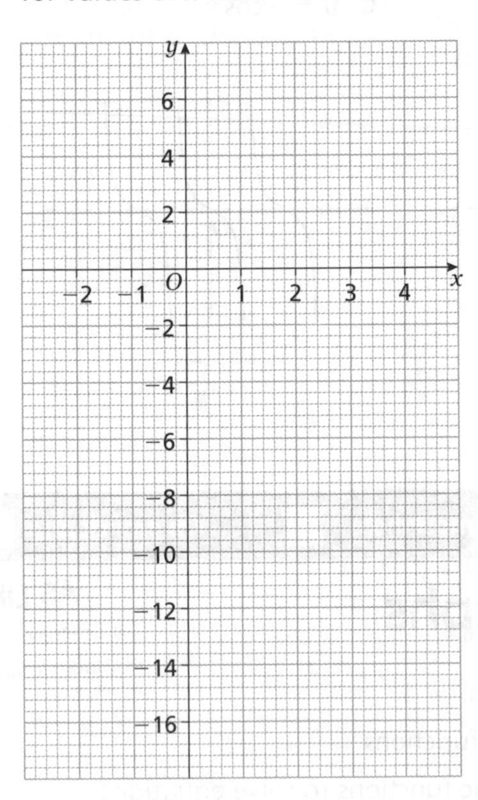

b Use your graph to find estimates for the solutions of $2x^2 - 4x = 3$.

$x \approx$ .... or $x \approx$ ....

3 a On the grid below, draw the graph of $y = x^2 + 5x - 3$ for values of x from -7 to 2.

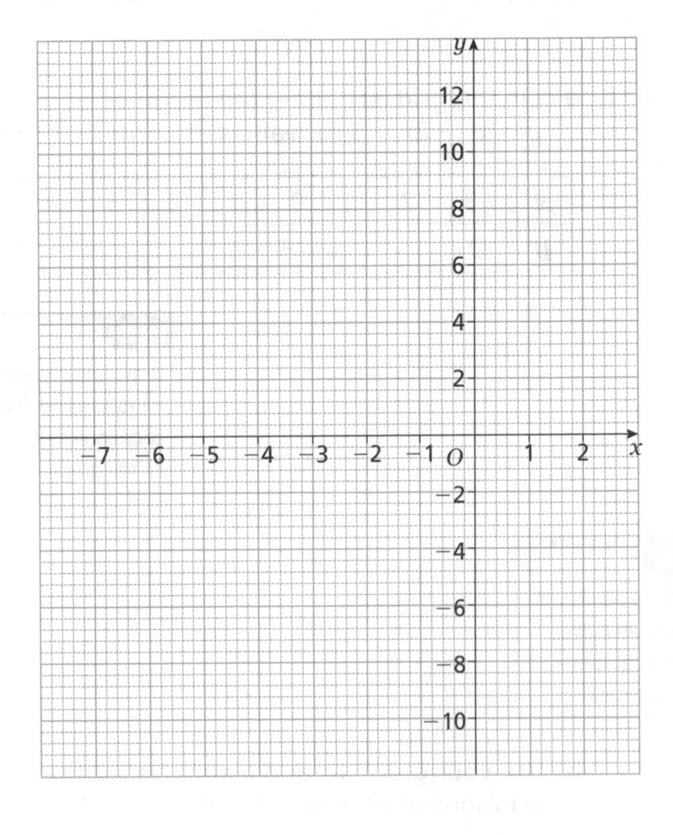

b Use your graph to find estimates for the solutions of $x^2 + 5x = 3$.

$x \approx$ .... or $x \approx$ ....

4 a On the grid below, draw the graph of $y = x^3 - 2x + 3$ for values of x from -3 to 3.

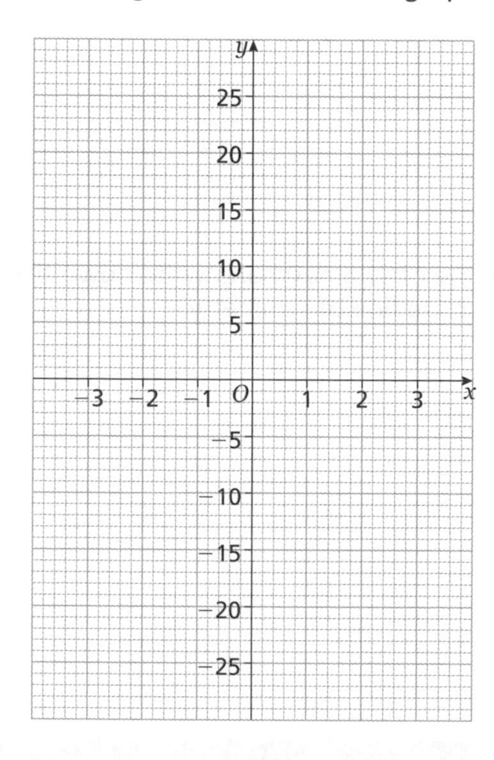

b Use your graph to find estimates for the solutions of $x^3 = 2x$.

Hint

Rewrite the equation $x^3 = 2x$ so that $x^3 - 2x + 3$ is on one side. The other side of the equation gives you the equation of the straight line which needs drawing on the graph.

$x =$ .... or $x \approx$ .... or $x \approx$ ....

5 a On the grid below, draw the graph of $y = 4 + 3x - x^2 - x^3$ for values of x from -3 to 3.

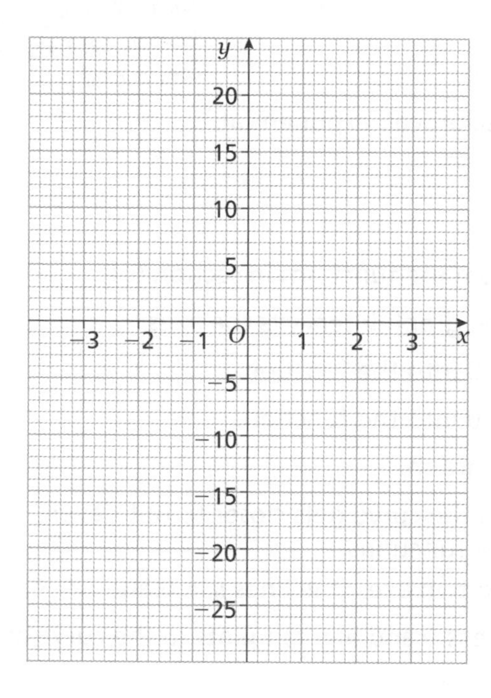

b Use your graph to find an estimate for the solution of $x^3 + x^2 = 3x + 4$.

$x \approx$ ....

c Use your graph to find estimates for the solutions of $4 = x^3 + x^2 - 4x$.

$x =$ .... or $x =$ .... or $x =$ ....

Step into AS

6 On the grid below, draw the graph of $y = x^2 - 4x$ for values of x from -1 to 5.

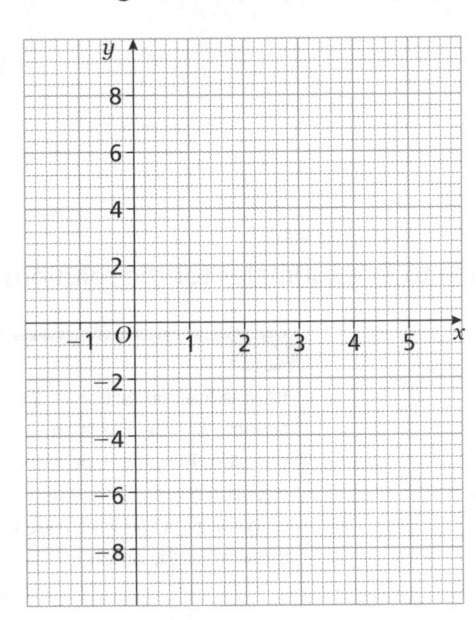

b Draw the line of symmetry and write down the equation of the line of symmetry.

....

Needs more practice ☐ Almost there ☐ I'm proficient! ☐

9.3 Sketching graphs

AS LINKS

C1: 4.3 Sketching the reciprocal function; **C2:** 8.4 Graphs of sin θ, cos θ and tan θ

By the end of this section you will know how to:

* Sketch graphs of linear, quadratic, cubic, reciprocal, exponential, and circular functions

Key points

* It is important to know the general shape of the graphs from the functions in section 9.1.
* To sketch the graph of a function, find the points where the graph intersects the axes.
* To find where the curve intersects the y-axis substitute $x = 0$ into the function.
* To find where the curve intersects the x-axis substitute $y = 0$ into the function.
* Where appropriate, mark and label the asymptotes on the graph.
* **Asymptotes** are lines (usually horizontal or vertical) which the curve gets closer to but never touches or crosses. Asymptotes usually occur with reciprocal functions.
  For example, the asymptotes for the graph of $y = \frac{a}{x}$ are the two axes (the lines $y = 0$ and $x = 0$).
* At the **turning points** of a graph the gradient of the curve is 0 and any tangents to the curve at these points are horizontal.
* To find the coordinates of the **maximum** or **minimum point** (turning points) of a quadratic curve (parabola) you can use completed square form.
* A **double root** is when two of the solutions are equal. For example, $(x - 3)^2(x + 2)$ has a double root at $x = 3$.
* When there is a double root, this is one of the turning points of a cubic function.

Guided

1 Sketch the graph of $y = x^2$.

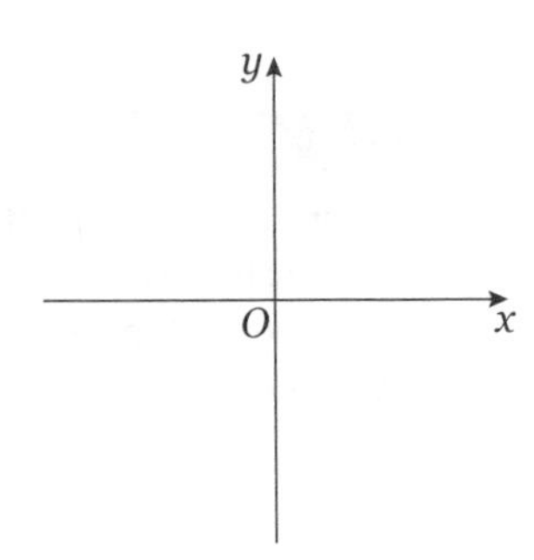

Hint
The graph of $y = x^2$ is a parabola (see section 9.1).

2 Sketch the graph of $y = x^2 - x - 6$.

To find where the graph intersects the axes:

When $x = 0$, $y = 0^2 - 0 - 6 =$ ......,

so the graph intersects the y-axis at the point (0, ...).

When $y = 0$, $x^2 - x - 6 = 0$

Factorising this equation gives:

$(x -$ ....$)(x +$ ....$) = 0$

$x =$ .... or $x =$ ....,

so the graph intersects the x-axis at the points (.........., 0) and (.........., 0).

This equation has a positive coefficient of x^2

(so for a quadratic in the form

$y = ax^2 + bx + c$, $a > 0$).

So the graph is a ⋁ shape (not a ⋀ shape).

For the turning point complete the square:

$$x^2 - x - 6 = \left(x - \frac{1}{2}\right)^2 - \frac{1}{4} - 6$$

$$= \left(x - \frac{1}{2}\right)^2 - \text{..........}$$

The turning point is the minimum value for this expression (this is when the term in the bracket is equal to zero).

When $\left(x - \frac{1}{2}\right)^2 = 0$, $x =$ .... and $y =$ ....,

so the turning point is at the point (.........., ...).

y
−2 O 3 x

Hint
Mark the points of intersection and the turning point clearly on the graph.

3 Sketch the graph of $y = \sin x$ for values of x from $-180°$ to $180°$.

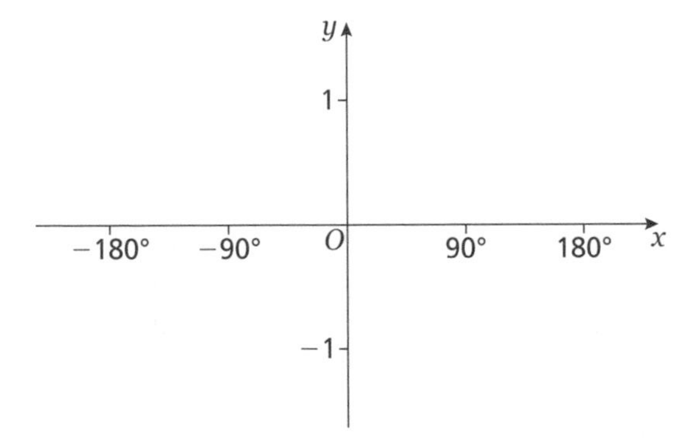

Hint
For the general shape of the graph $y = \sin x$ look at the Key points for section 9.1.

4 Sketch the graph of $y = (x - 3)(x - 1)(x + 2)$.

When $y = 0$, $0 = (x - 3)(x - 1)(x + 2)$

So $x - 3 = 0$, $x - 1 = 0$ or $x + 2 = 0$

$x =$ ...., ... or ....

so the graph intersects the x-axis at the points

(.........., 0), (.........., 0) and (.........., 0).

When $x = 0$, $y = (0 - 3)(0 - 1)(0 + 2)$

$=$ .... $\times$ .... $\times$ .... $=$ ....

so the graph intersects the y-axis at the point (0, ...).

Hint

For the general shape of the graph look at the Key points for section 9.1 (the cubic graph for $a > 0$).

5 Sketch the graph of $y = (x + 2)^2(x - 1)$.

The solutions for this function are $x = -2$ or $x = 1$. As $x = -2$ is the 'double root', this is the turning point which can be identified.

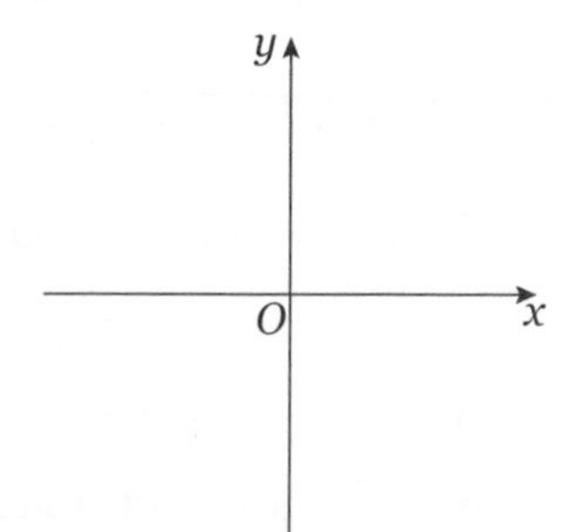

6 Sketch the graph of $y = -x^2$.

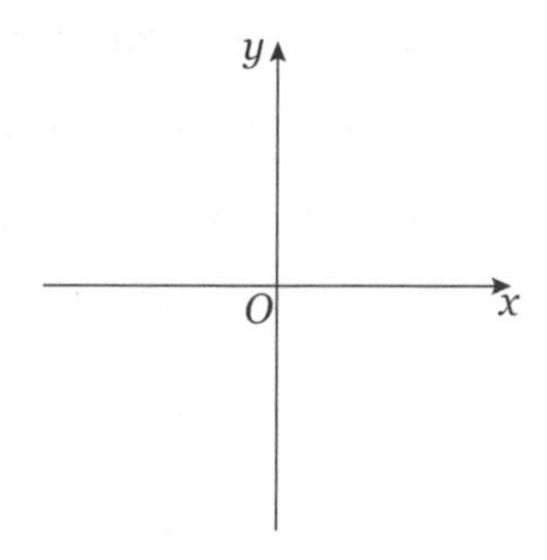

7 Sketch the graph of $y = x^2 - 5x + 6$.

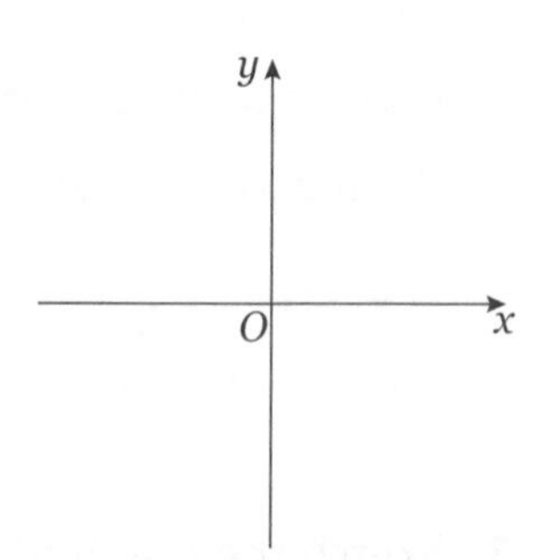

8 Sketch the graph of $y = -x^2 + 7x - 12$.

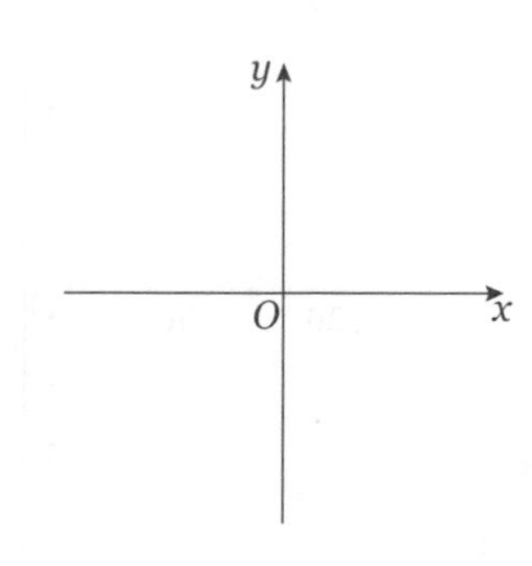

Practice

9 Sketch the graph of $y = -x^2 + 4x$.

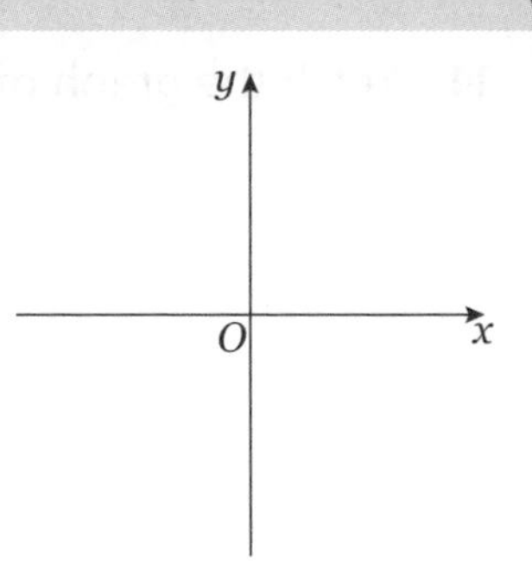

10 Sketch the graph of $y = x^2 + 2x + 1$.

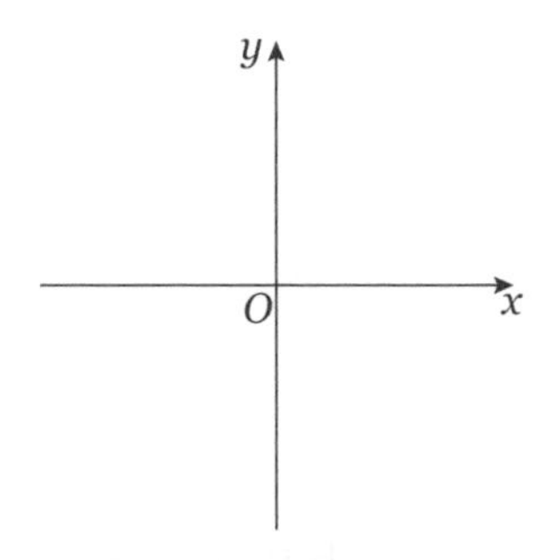

11 Sketch the graph of $y = 2x^3$.

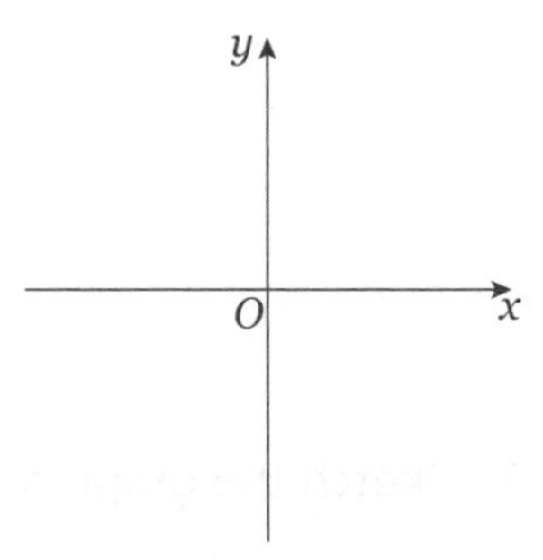

12 Sketch the graph of $y = x(x - 2)(x + 2)$.

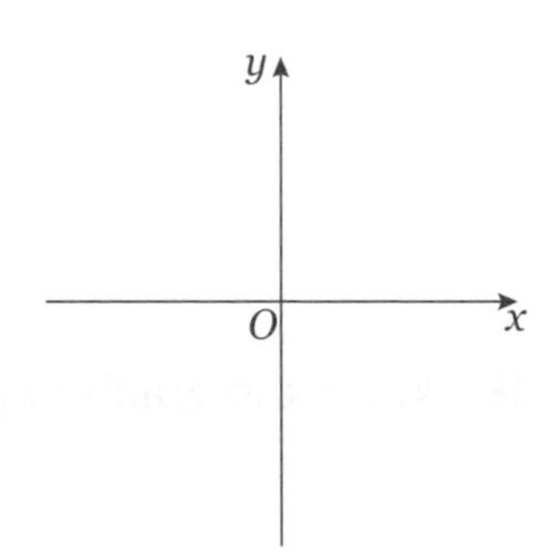

13 Sketch the graph of $y = (x + 1)(x + 4)(x - 3)$.

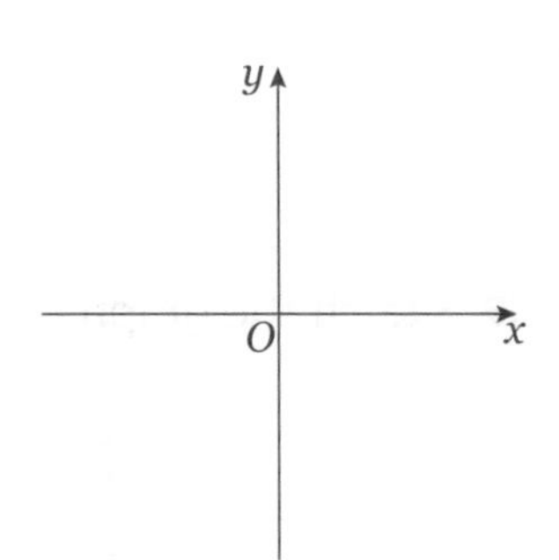

14 Sketch the graph of $y = (x + 1)(x - 2)(1 - x)$.

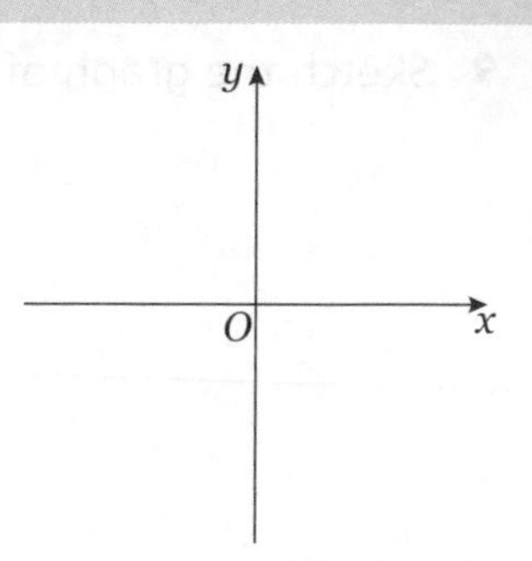

15 Sketch the graph of $y = (x - 3)^2(x + 1)$.

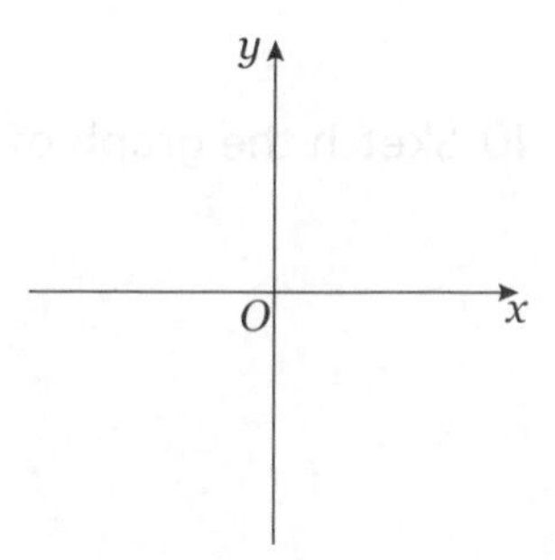

16 Sketch the graph of $y = (x - 1)^2(x - 2)$.

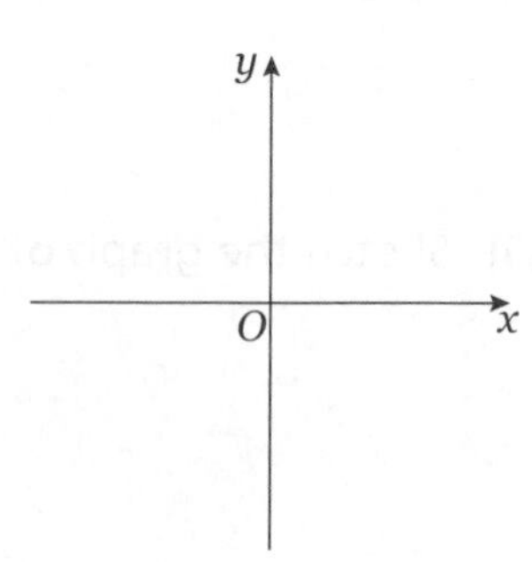

17 Sketch the graph of $y = 3^x$.

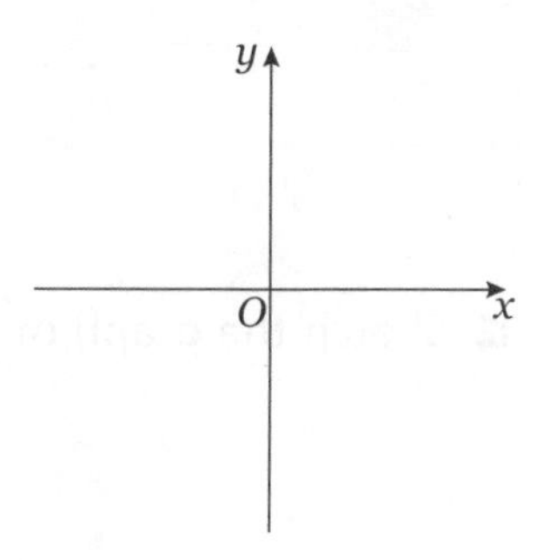

18 Sketch the graph of $y = \frac{3}{x}$.

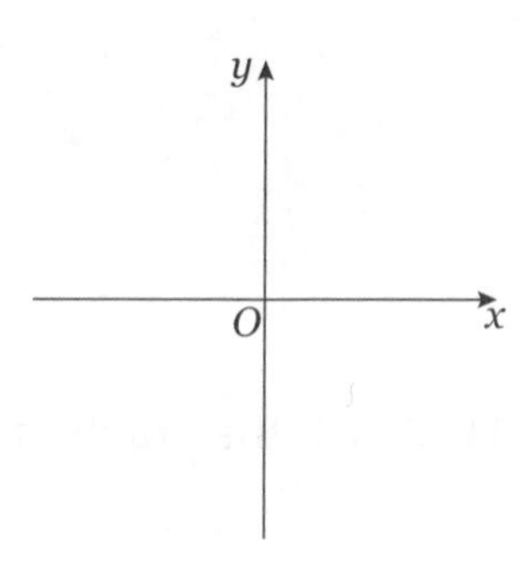

19 Sketch the graph of $\frac{1}{x + 2}$.

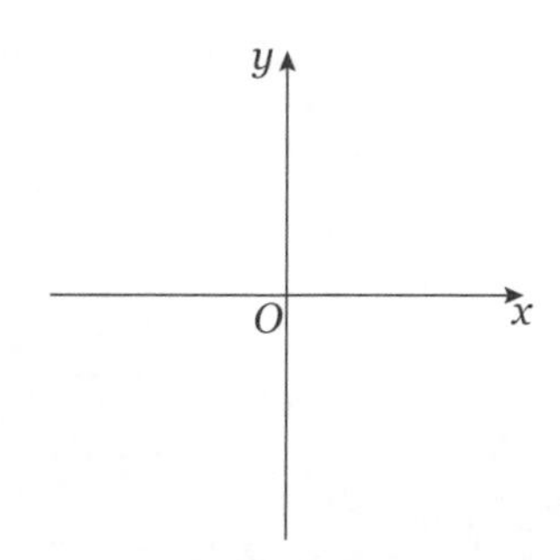

20 Sketch the graph of $\frac{1}{x - 1}$.

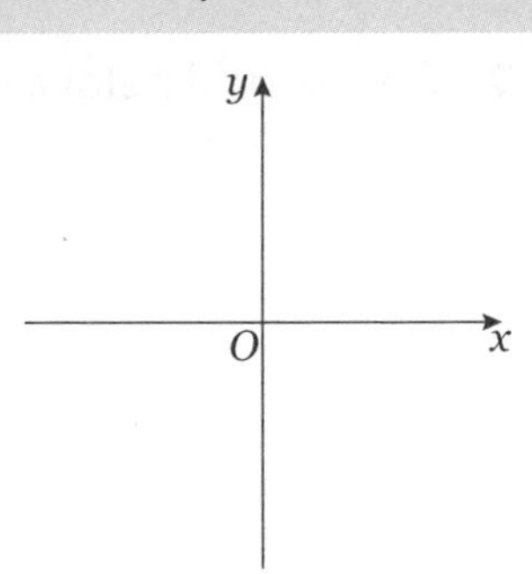

21 Sketch the graph of $y = \cos x$.

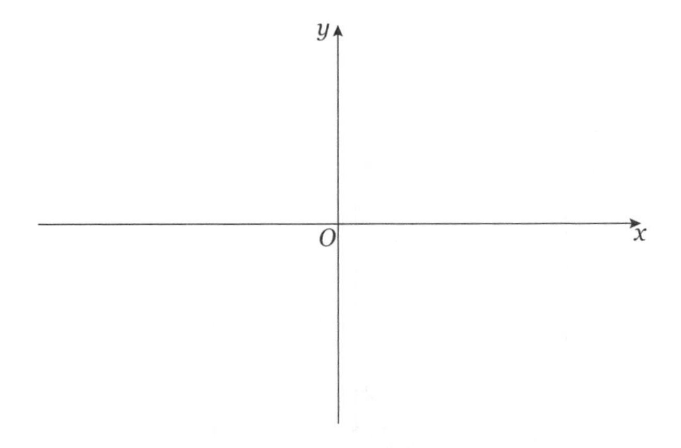

Needs more practice ☐ Almost there ☐ I'm proficient! ☐

9.4 Graphs of circles

AS LINKS
C2: 4.3 The equation of a circle

By the end of this section you will know how to:

- Construct circles using their equations

Key points

- A circle with centre at (0, 0) and radius r has an equation of the form $x^2 + y^2 = r^2$.
- The equation of a circle is in the form $(x - a)^2 + (y - b)^2 = r^2$ where the point (a, b) is the centre of the circle and r is the radius of the circle.

Guided

1 On the grid below, draw the graph of $x^2 + y^2 = 16$.

$\sqrt{16} = 4$

so the radius of the circle is 4

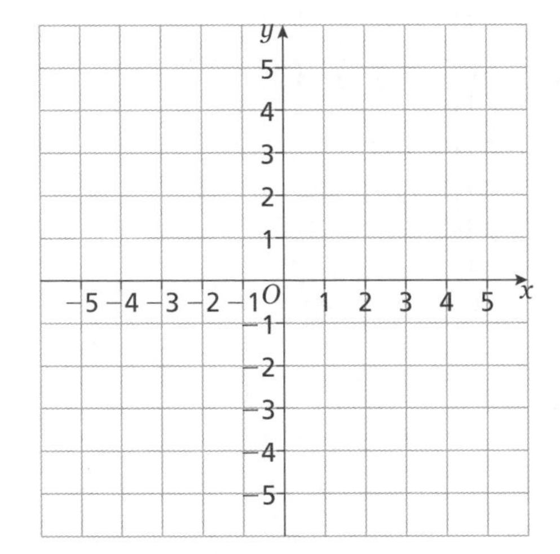

2 On the grid below, draw the graph of $(x - 1)^2 + (y + 1)^2 = 9$.

$\sqrt{9} = 3$

so the radius of the circle is 3

Hint
$a = 1$, $b = -1$, so the centre of the circle is at (1, −1).

Practice

3 On the grid below, draw the graph of $x^2 + y^2 = 25$.

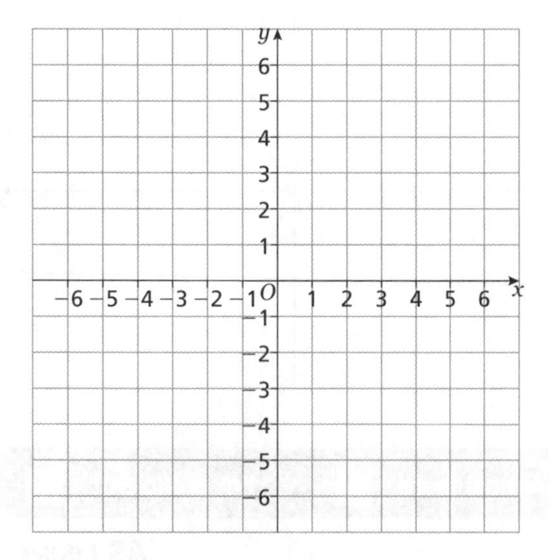

4 On the grid below, draw the graph of $x^2 + y^2 = 4$.

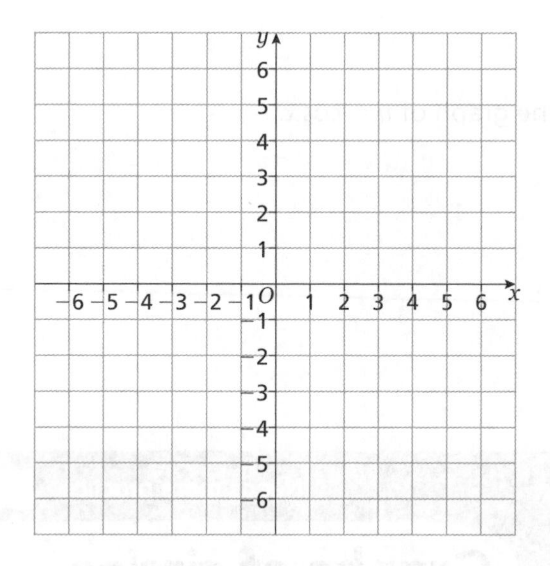

5 On the grid below, draw the graph of $x^2 + y^2 = 9$.

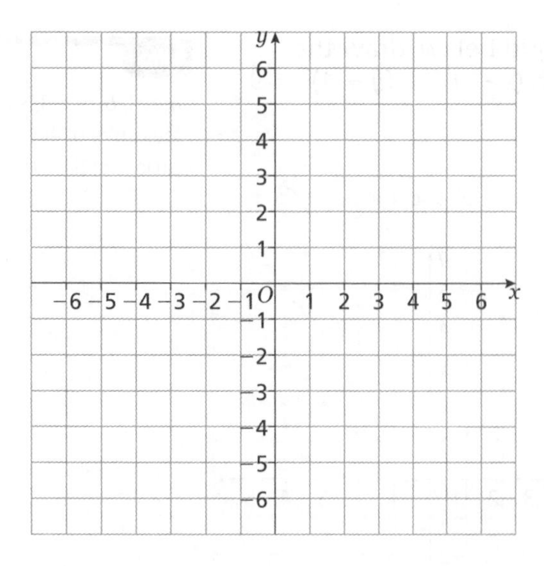

6 On the grid below, draw the graph of $x^2 + y^2 = 36$.

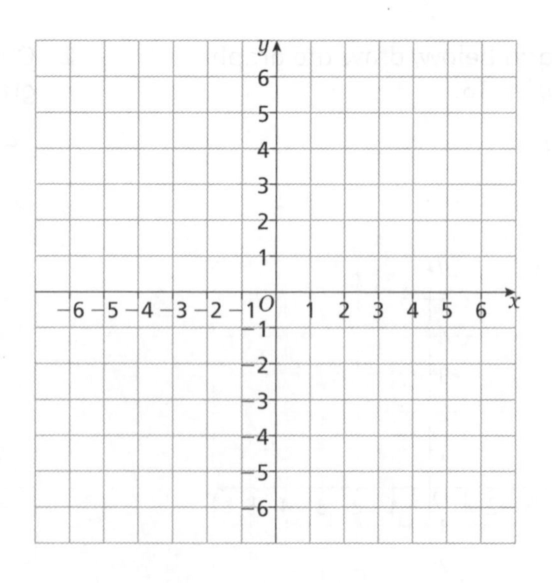

7 The equation of a circle is $(x - 1)^2 + (y + 2)^2 = 9$.

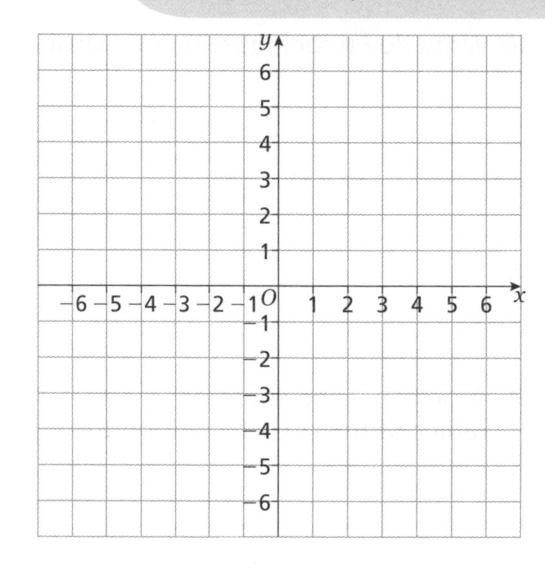

a Write down the coordinates of the centre of the circle.

(.........., ...)

b Write down the radius of the circle. ....

c On the grid, draw the circle with equation

$(x - 1)^2 + (y + 2)^2 = 9$.

8 The equation of a circle is $(x + 2)^2 + (y - 3)^2 = 4$.

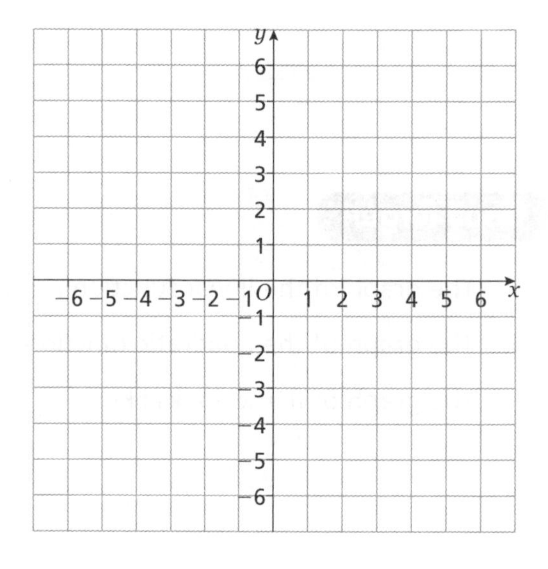

a Write down the coordinates of the centre of the circle.

(.........., ...)

b Write down the radius of the circle. ....

c On the grid, draw the circle with equation

$(x + 2)^2 + (y - 3)^2 = 4$.

9 The equation of a circle is $x^2 + (y - 2)^2 = 4$.

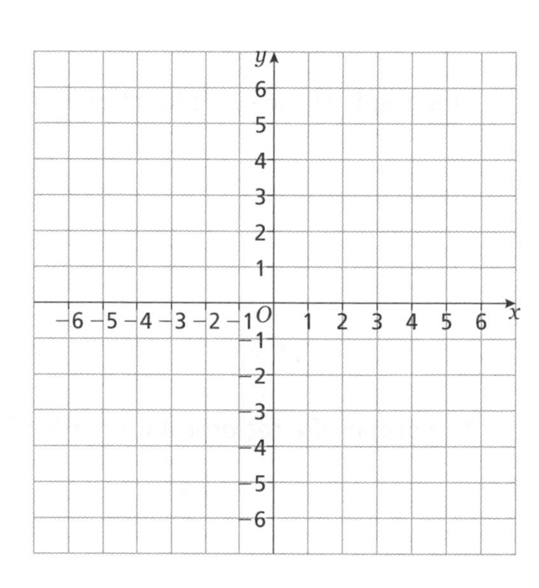

a Write down the coordinates of the centre of the circle.

(.........., ...)

b Write down the radius of the circle. ....

c On the grid, draw the circle with equation

$x^2 + (y - 2)^2 = 4$.

10 Write down the equation of a circle with centre (2, 3) and radius 6.

11 A circle has the equation $x^2 + y^2 + 4x - 10y + 13 = 0$. Find the coordinates of the centre and the radius of the circle by writing the equation in the form $(x - a)^2 + (y - b)^2 = r^2$.

Don't forget!

* The graph of the linear function $y = mx + c$ is ....
* The graph of the quadratic function $ax^2 + bx + c$, where $a \neq 0$, is a curve called a ....
* The graph of $y = x^2$ looks like:

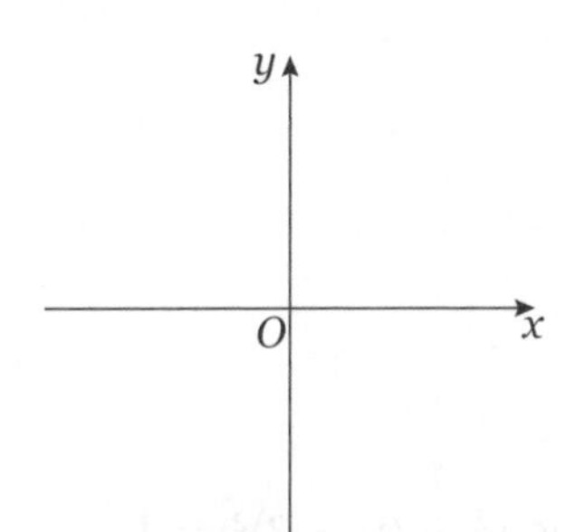

* The graph of the cubic function $y = x^3$ looks like:

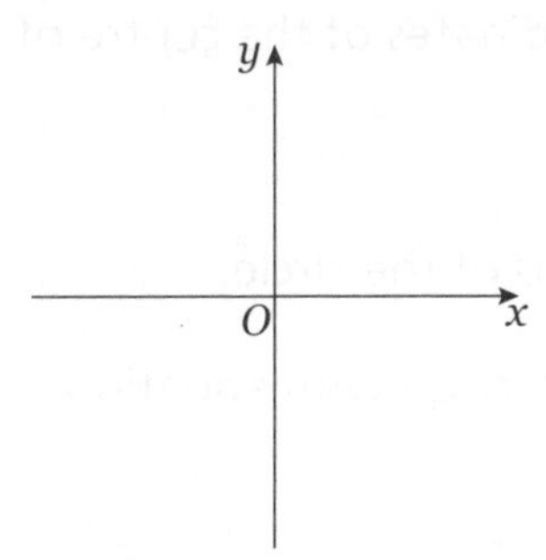

* The graph of a reciprocal function, of the form $y = \frac{a}{x}$, has these shapes:

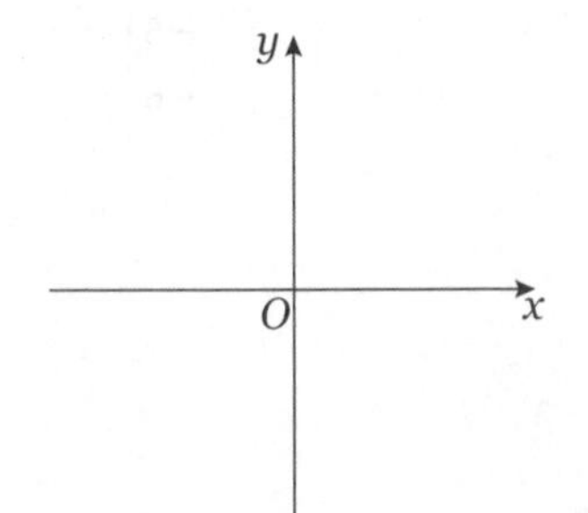

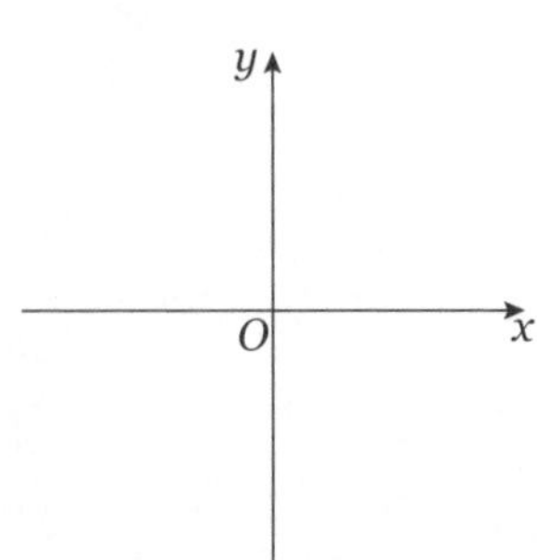

* The graph of an exponential function, of the form $y = a^x$, has these shapes:

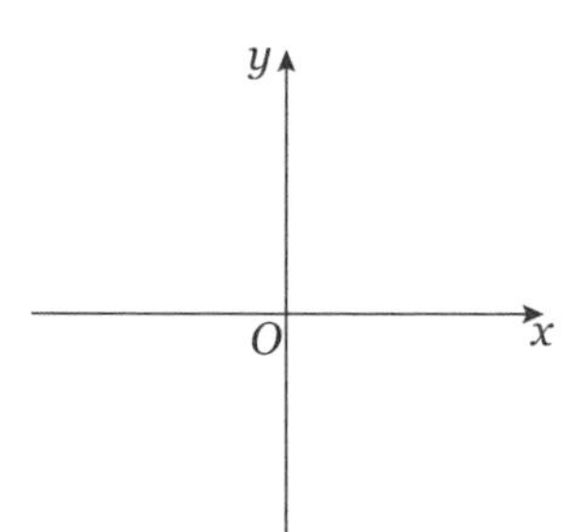

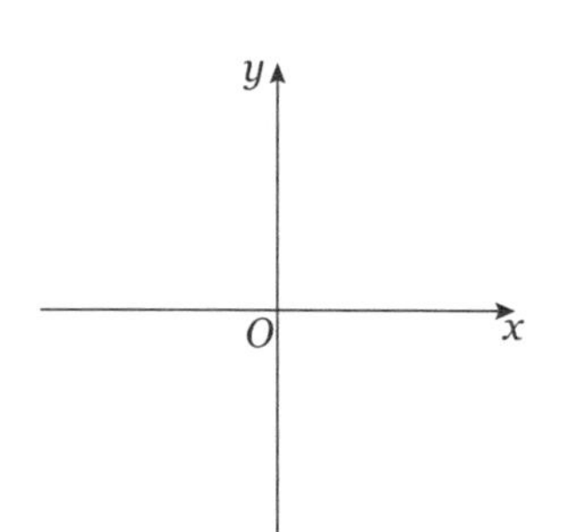

* Circular (or trigonometric) functions include sine, cosine or tangent. The graph of a circular function has one of these shapes:

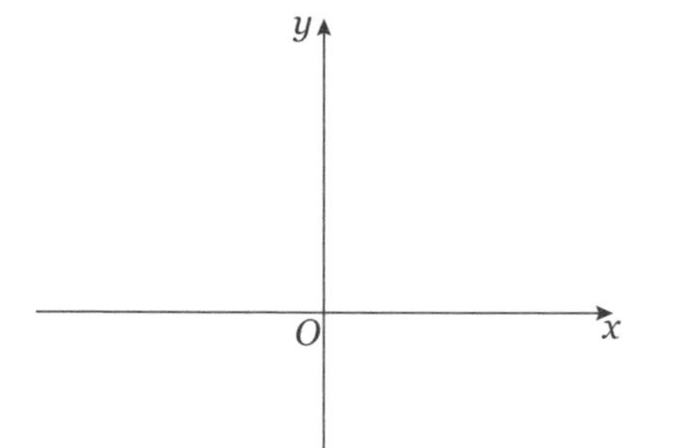

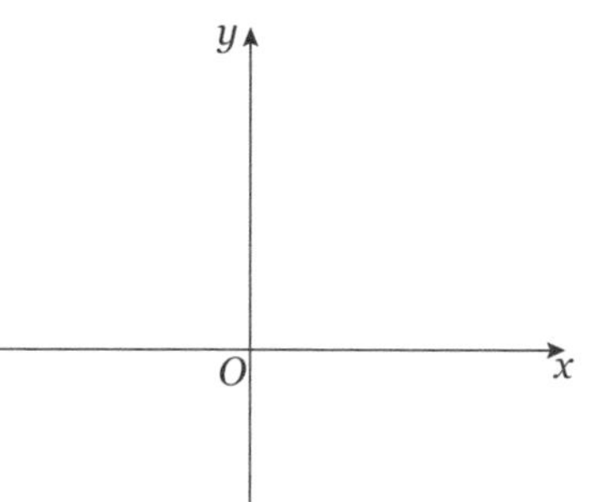

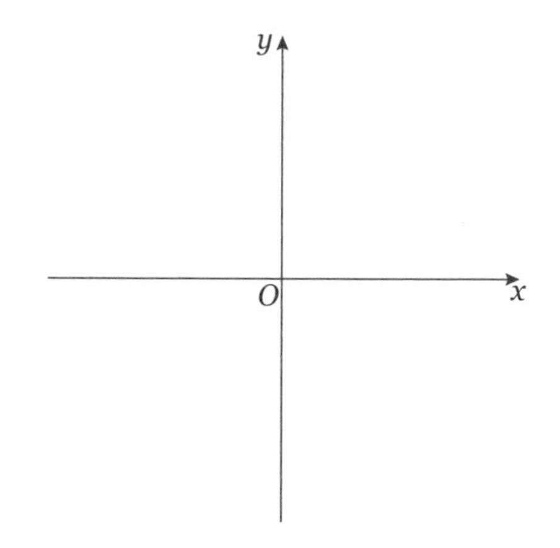

* The tangent to a curve is a straight line which ....

....

* The normal to a curve is .... to the tangent at that point on the curve.

* When sketching a curve to find where the curve intersects the y-axis substitute .... into the function.

* When sketching a curve to find where the curve intersects the x-axis substitute .... into the function.

* Asymptotes are lines (usually horizontal or vertical) which ....

....

* To calculate the turning point of a quadratic function, ... to find the minimum or maximum value.

* When there is a 'double root', this is one of the .... of a cubic function.

* The equation of a circle is in the form .... where the point (a, b) is the .... of the circle and r is the .... of the circle. If $a = 0$ and $b = 0$, then the equation is of the form .... and the centre of the circle is at (0, 0).

Exam-style questions

1 Here are some sketch graphs.

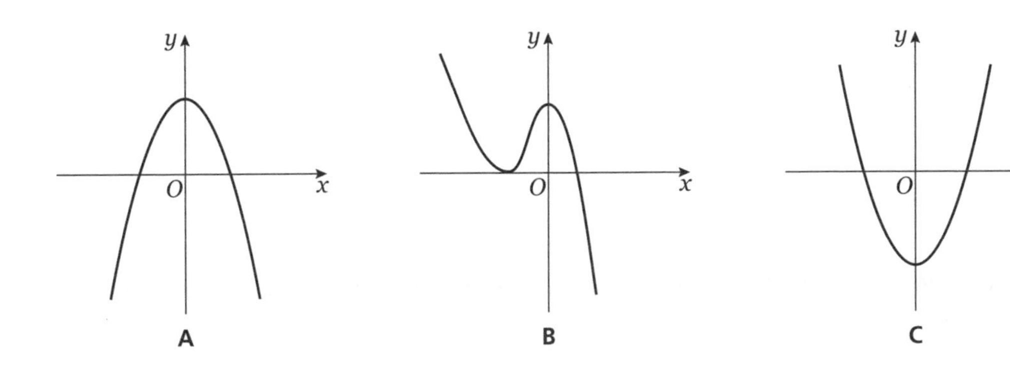

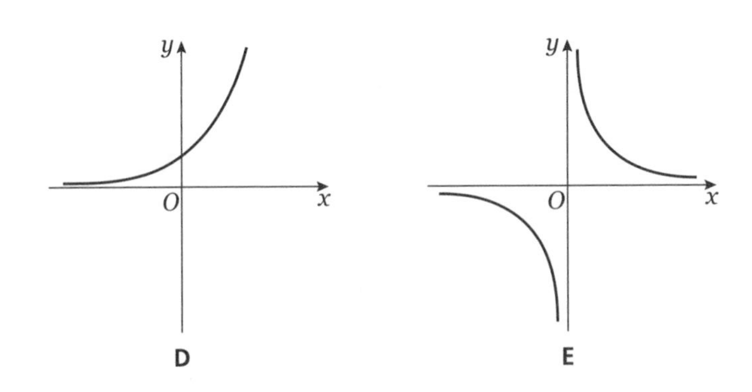

The table shows the equations of some graphs.

Equation	**Graph**
$y = 3^x$	
$y = (x + 2)(x - 2)$	
$y = (2 - x)(2 + x)$	
$y = \frac{2}{x}$	
$y = (x + 2)^2(1 - x)$	

Match the letter of the graph with its equation.

2 Sketch the graph of $y = (x + 1)(x - 2)(x - 3)$

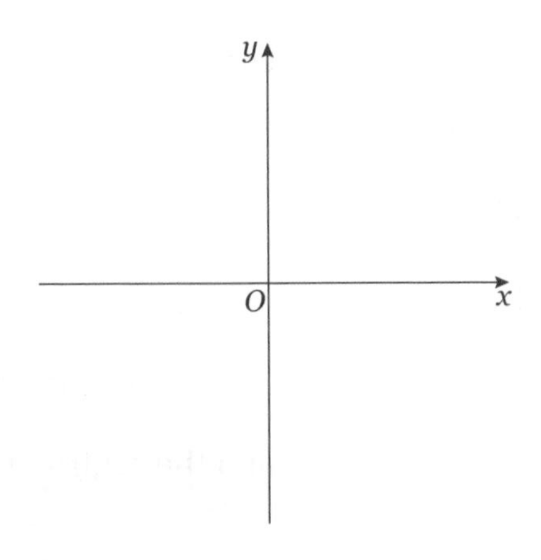

3 a On the grid below, draw the graph of $y = x^3 - x^2 - 6x$ for values of x from -3 to $+4$

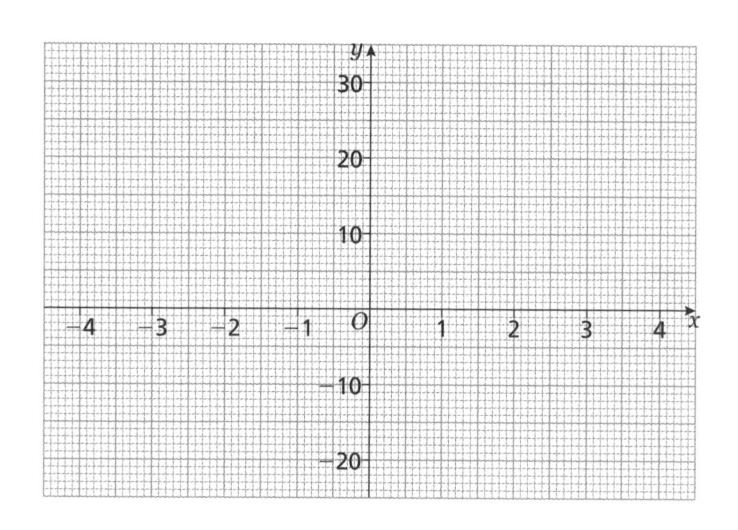

b Use your graph to find estimates for the solutions of $x^3 - x^2 - 6x = -5$

....

4 On the grid below, draw the graph of $x^2 + y^2 = 9$

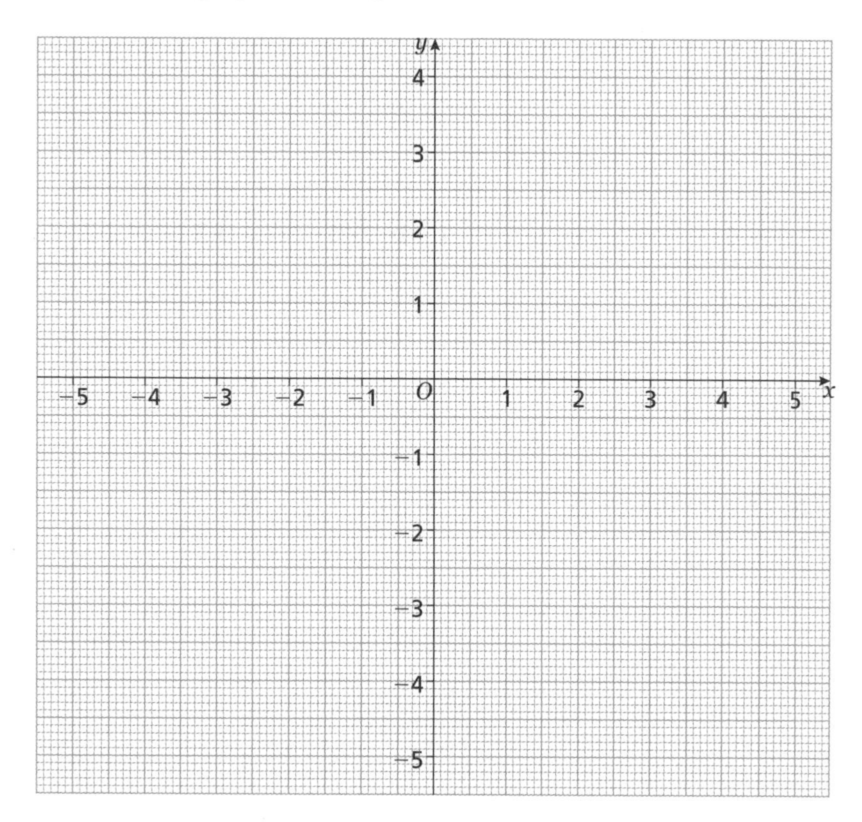

10.1 Solving linear inequalities

AS LINKS
C1: 3.4 Solving linear inequalities

By the end of this section you will know how to:

* Solve linear inequalities

Key points

* Solving linear inequalities uses similar methods to those for solving linear equations.
* When you multiply or divide an inequality by a negative number you need to reverse the inequality sign, e.g $<$ becomes $>$.

Guided

1 Solve

a $-8 \leqslant 4x < 16$

.... $\leqslant x <$ ....

Hint
Divide all three terms by 4.

b $4 < 5x \leqslant 10$

.... $< x \leqslant$ ....

c $2x - 5 < 7$

$2x <$ ....

$x <$ ....

d $2 - 5x \geqslant -8$

$-5x \geqslant$ ....

x ....

e $4(x - 2) > 3(9 - x)$

$4x - 8 >$ ....

.... $>$ ....

$x >$ ....

Practice

2 Solve

a $2 - 4x \geqslant 18$

b $3 \leqslant 7x + 10 < 45$

c $6 - 2x \geqslant 4$

d $4x + 17 < 2 - x$

e $4 - 5x < -3x$

f $-4x \geqslant 24$

Step into AS

3 Solve

a $3(2 - x) > 2(4 - x) + 4$

b $5(4 - x) > 3(5 - x) + 2$

4 Find the set of values of x for which $2x + 1 > 11$ and $4x - 2 > 16 - 2x$.

Needs more practice ☐ Almost there ☐ I'm proficient! ☐

10.2 Solving quadratic inequalities

By the end of this section you will know how to:

* Solve quadratic inequalities

AS LINKS
C1: 3.5 Solving quadratic inequalities

Key points

* First solve the quadratic equation.
* Sketch the graph of the quadratic function.
* Use the graph to find the values which satisfy the quadratic inequality.

Guided

Find the set of values of x which satisfy the following inequalities.

1 $x^2 + 5x + 6 > 0$

Solve $x^2 + 5x + 6 = 0$

$(x$ .... $)(x$ .... $) = 0$

$x =$ ...., $x =$ ....

The values which satisfy the inequality

$x^2 + 5x + 6 > 0$ are

$x <$ .... or $x >$ ....

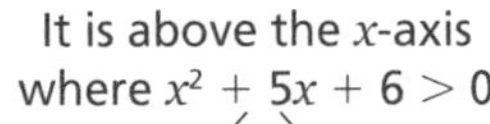

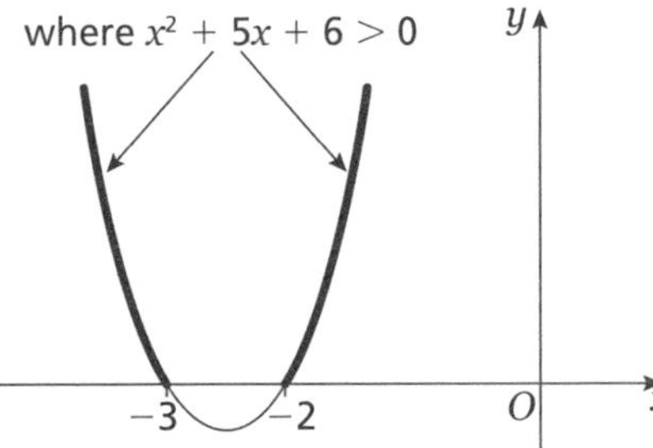

Hint

On the graph, identify where the graph of $x^2 + 5x + 6 > 0$ lies above the x-axis, i.e. where $y > 0$.

2 $x^2 - 5x \leqslant 0$

Solve $x^2 - 5x = 0$

$x($ .... $) = 0$

$x =$ ...., $x =$ ....

The range which satisfies the inequality is .... $\leqslant x \leqslant$ ....

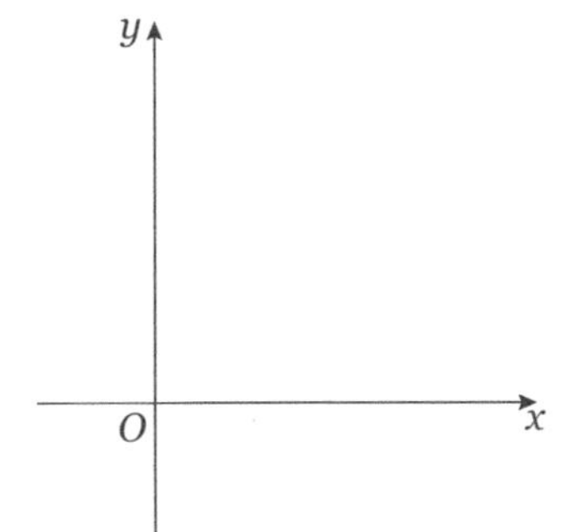

3 $-x^2 - 3x + 10 \geqslant 0$

Solve $-x^2 - 3x + 10 = 0$

Rearranging this gives x^2 ..... $3x$ ..... $10 = 0$

(..............)(..............) = 0

$x =$ ...., $x =$ ....

The values which satisfy the inequality are .... $\leqslant x \leqslant$ ....

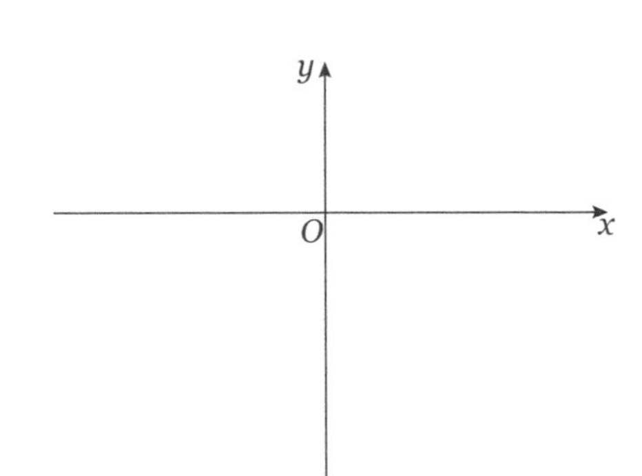

Practice

4 Find the set of values of x for which $x^2 - 4x - 12 \geqslant 0$.

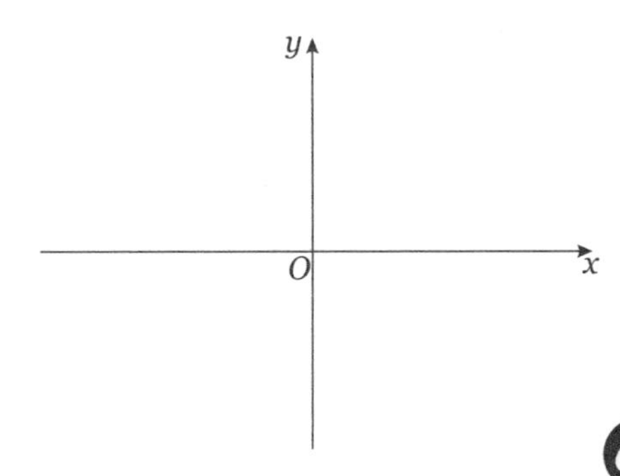

5 Find the set of values of x for which $(x + 7)(x - 4) \leqslant 0$.

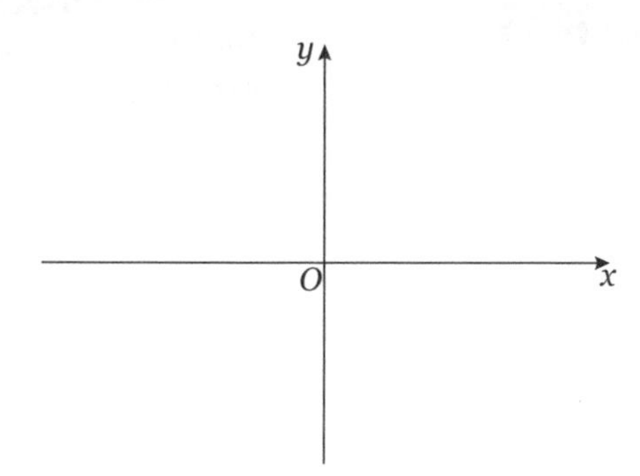

6 Find the set of values of x for which $2x^2 - 7x + 3 < 0$.

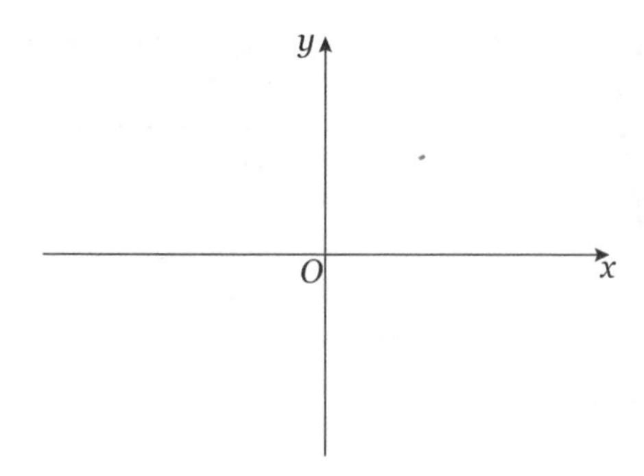

7 Find the set of values of x for which $4x^2 + 4x - 3 > 0$.

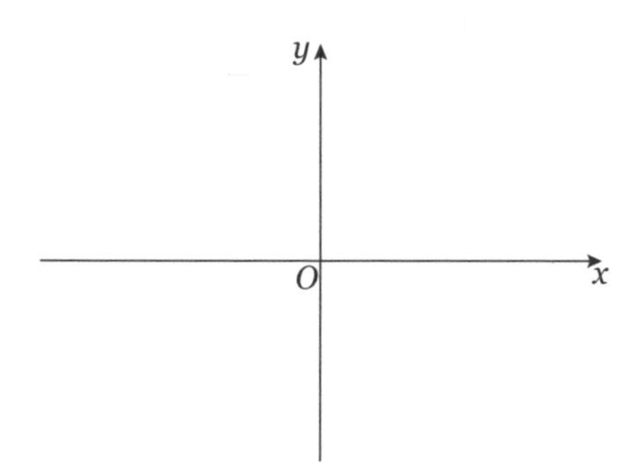

8 Find the set of values of x for which $12 + x - x^2 \geqslant 0$.

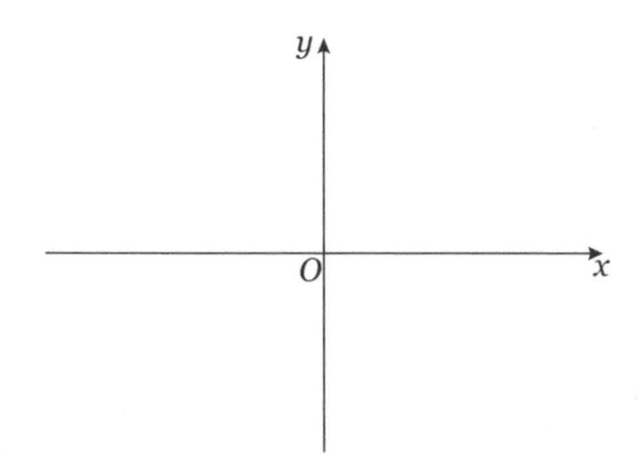

Find the set of values which satisfy the following inequalities.

9 $x(2x - 9) < -10$

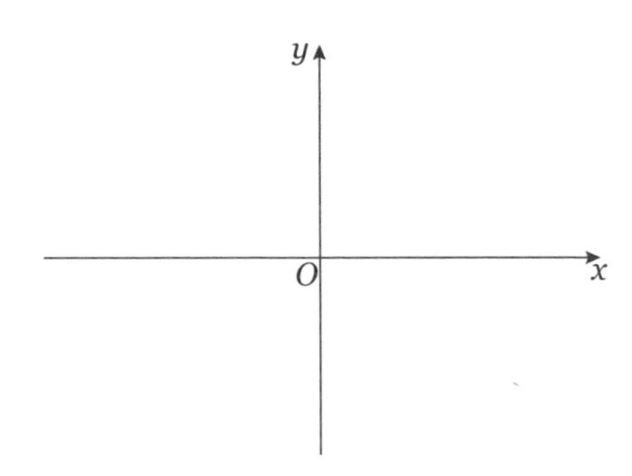

10 $6x^2 \geqslant 15 + x$

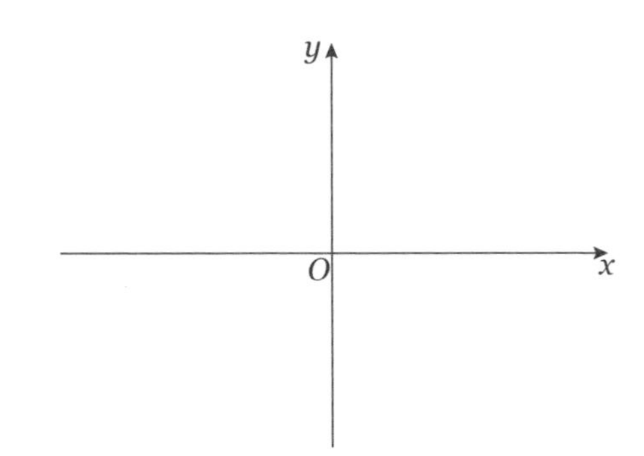

Needs more practice ☐ Almost there ☐ I'm proficient! ☐

10.3 Representing linear inequalities on a graph

AS LINKS

D1: 6.2 Illustrating a two-variable linear programming problem graphically

By the end of this section you will know how to:

- Represent linear inequalities on a graph

Key points

- Inequalities can be represented on a graph by shading regions.
- Inequalities which include $\leqslant$ or $\geqslant$ signs are shown using unbroken (solid) lines.
- Inequalities which include $<$ or $>$ signs are shown using broken lines.

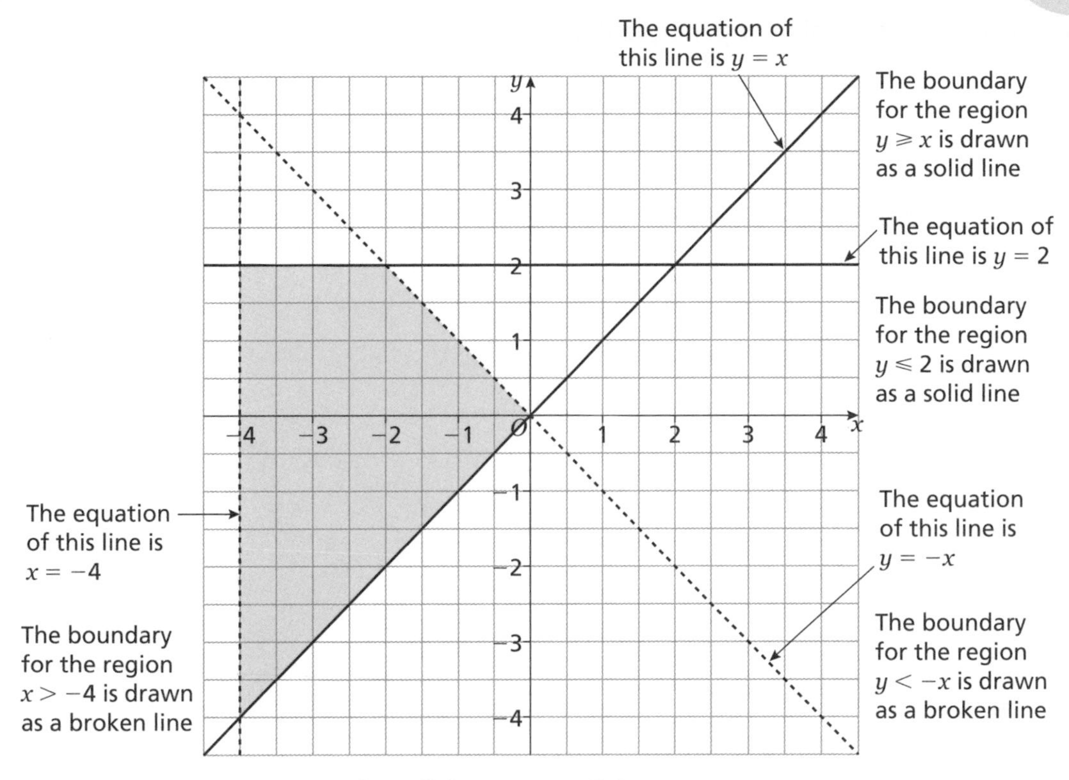

The shaded region satisfies all four inequalities: $x > -4$, $y \geqslant x$, $y \leqslant 2$, $y < -x$

Guided

1 On the grid, shade the region that satisfies the inequality $x > -3$

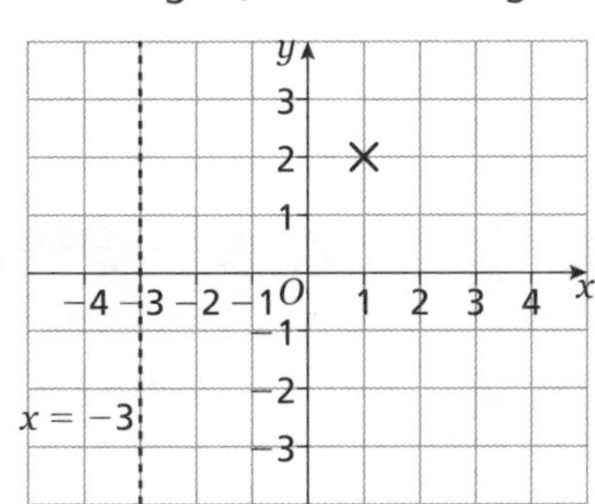

To check which region to shade in, choose a coordinate to test with the inequality $x > -3$. For this question, the point at (1, 2) has been chosen. As the x-coordinate is 1 we test this; as $1 > -3$ we know that the right-hand side of the broken line needs to be shaded.

2 On the grid, shade the region that satisfies all three of the inequalities.

$x \leqslant 3$

$y < x$

$y > -2$

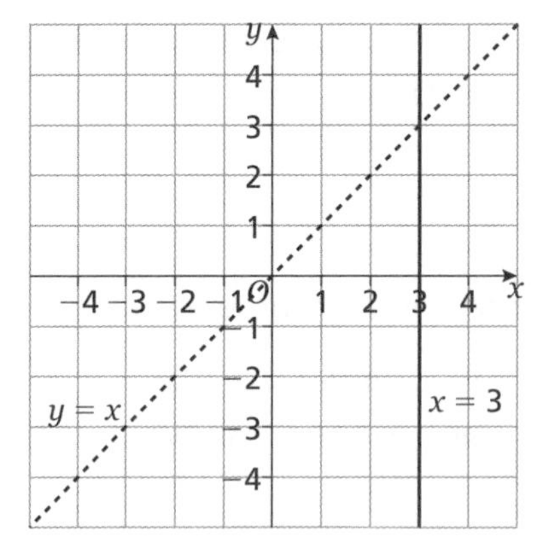

Hint

Test a coordinate for each inequality and then shade the region where all three inequalities are satisfied.

3 On the grid, shade the region that satisfies all three of the inequalities.

$x + y < 5$

$y \leqslant 2x + 3$

$y \geqslant 0$

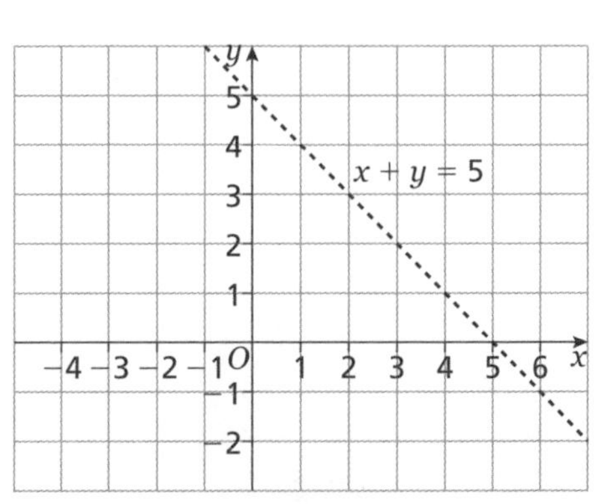

Practice

4 On the grid, shade the region that satisfies all three of the inequalities.

$y \leqslant 4$

$y > -x$

$y \geqslant 3x - 4$

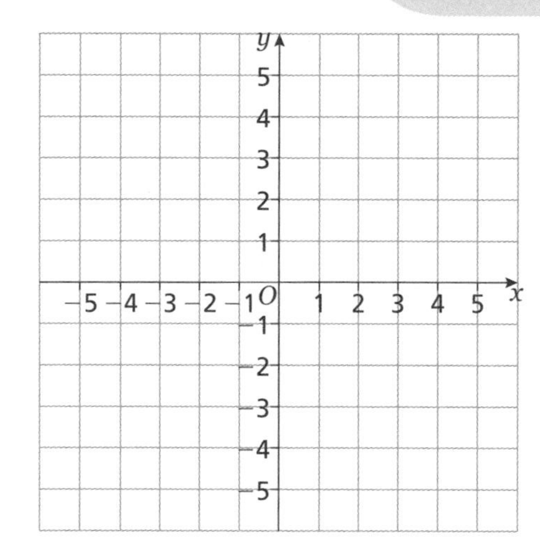

5 On the grid, shade the region that satisfies all three of the inequalities.

$y < 4$

$x > -3$

$y \geqslant \frac{1}{2}x$

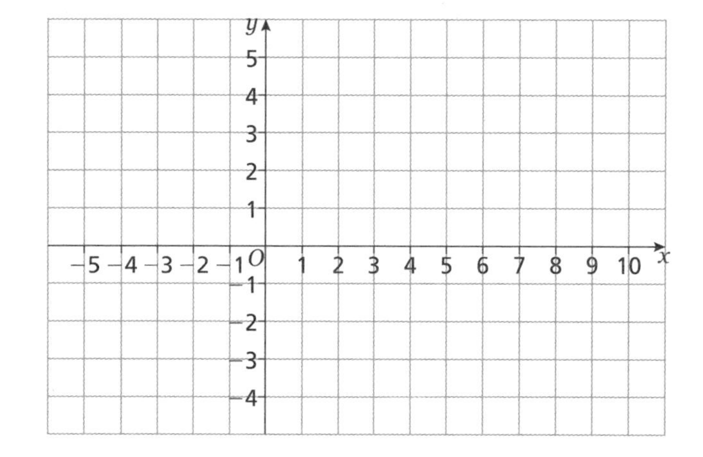

6 On the grid, shade the region that satisfies all three of the inequalities.

$y - x \leqslant 4$

$y > -3$

$x + y \leqslant 2$

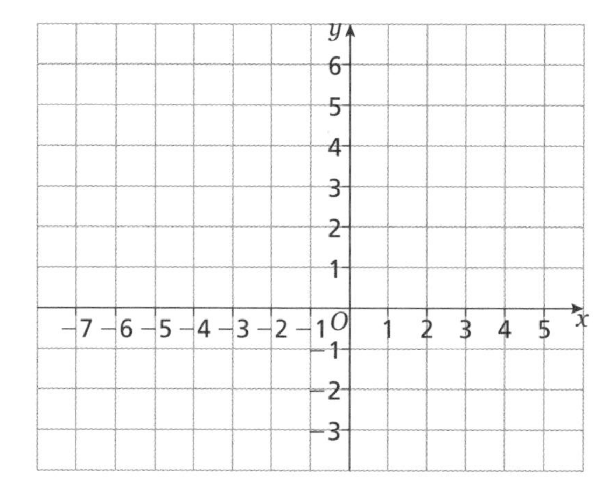

Step into AS

7 On the grid, shade the region R, that satisfies all four of the inequalities.

$y \leqslant 2x$

$x + y \leqslant 6$

$y > x - 3$

$y > -x$

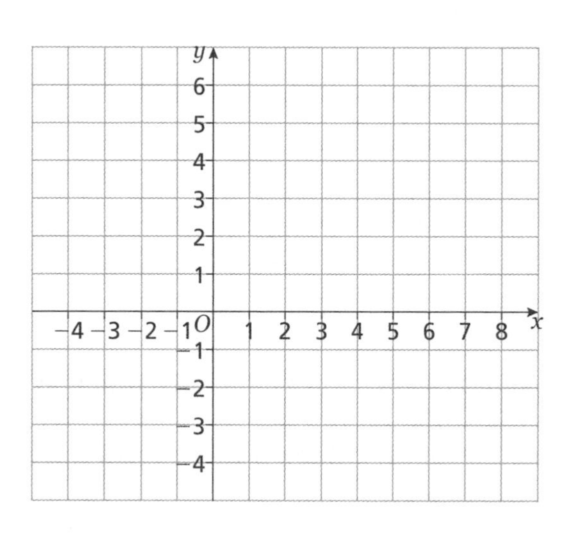

8 On the grid, shade the region that satisfies all three of the inequalities.

$y > -3$

$2x + 5y \leqslant 4$

$y \leqslant 3x + 2$

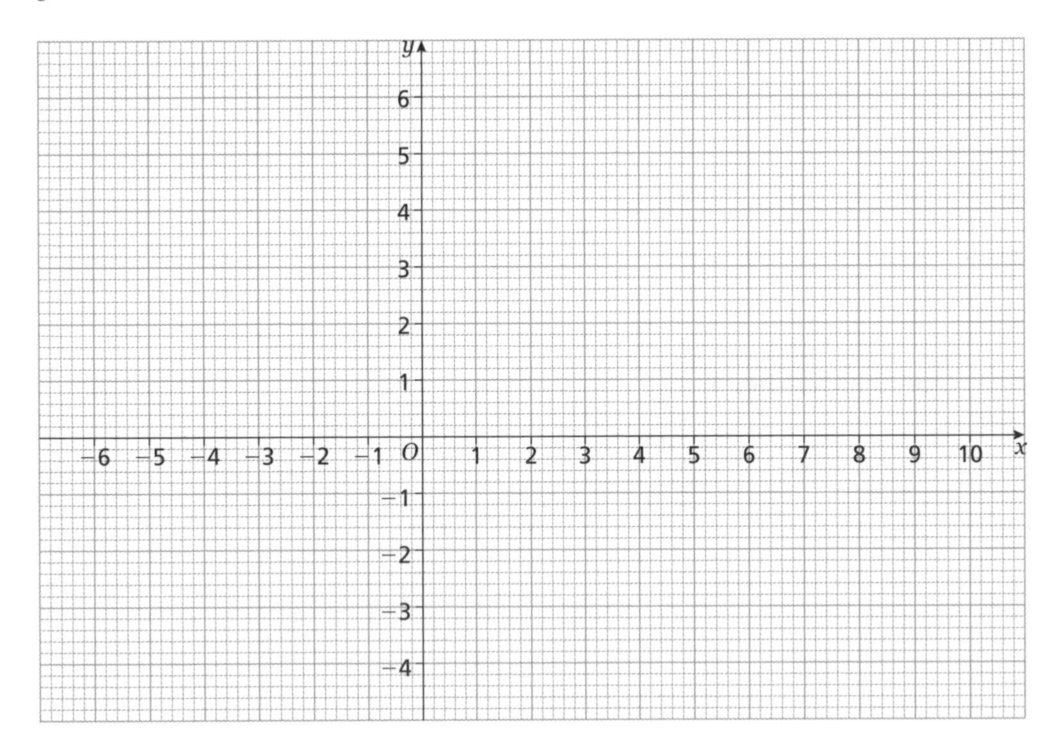

Don't forget!

* When you multiply or divide an inequality by .... you need to reverse the inequality sign.
* To solve quadratic inequalities, first .... the quadratic equation, then .... of the quadratic function and, finally, use the graph to find the .... which satisfy the quadratic inequality.
* Inequalities can be represented on a graph by ....
* Inequalities which include $\leqslant$ or $\geqslant$ signs are shown using .... lines.
* Inequalities which include $<$ or $>$ signs are shown using .... lines.

Exam-style questions

1 Solve $x^2 + x \leqslant 6$

....

2 On the grid, shade the region which satisfies all these inequalities

$x \leqslant 4$

$x + y > -1$

$y < 2x - 3$

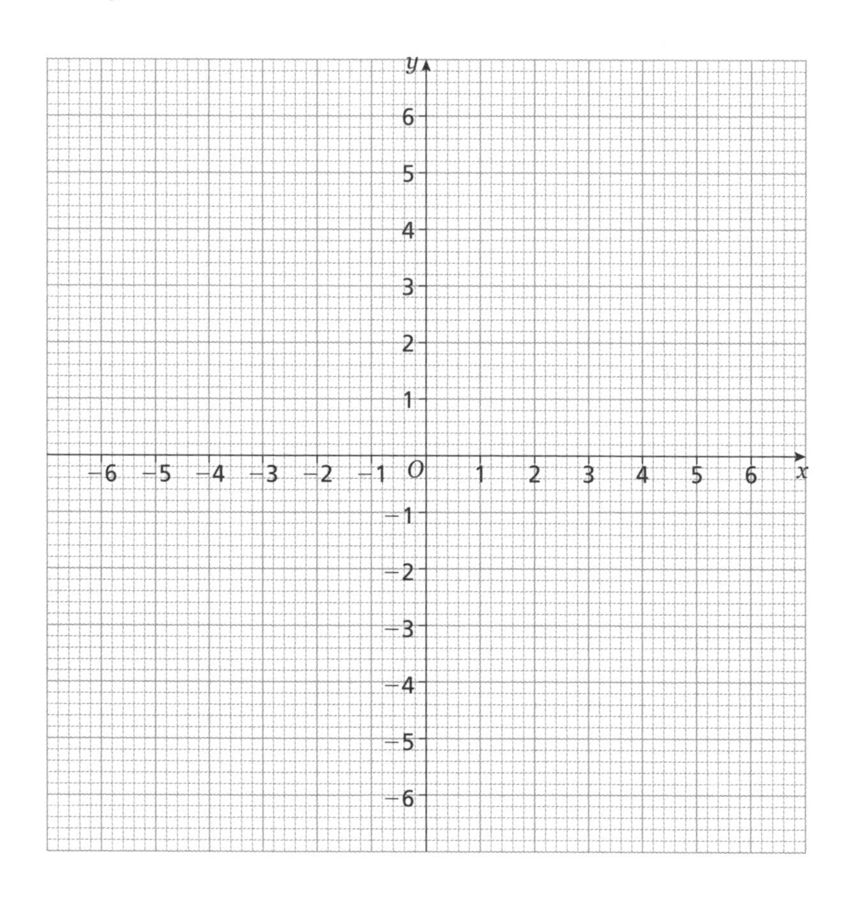

Needs more practice ☐ Almost there ☐ I'm proficient! ☐

11.1 Distance–time graphs

AS LINKS
M1: 2.4 Representing the motion of an object on a speed–time or distance–time graph

By the end of this section you will know how to:

* Draw, interpret and understand distance–time graphs

Key points

* On a distance–time graph the y-axis represents the distance travelled away from a starting point.
* On a distance–time graph the x-axis represents the time taken to travel.
* The gradient of a line on a distance–time graph represents speed; a straight line indicates constant speed. The steeper the line the faster the speed.
* Horizontal sections represent no movement.

Guided

1 The graph represents the movement of a lift in a block of flats.

Distance (metres) vs Time (seconds): axis values 0, 3, 6, 9, 12, 15, 18, 21 and 0, 10, 20, 30, 40, 50, 60, 70

a How many times in total did the lift stop? ....

b What was the speed of the lift between 30 and 40 seconds?

.... ÷ 10 = ... m/s

Hint
Calculating speed is the same as calculating the gradient. The gradient is calculated using the formula $m = \frac{y_2 - y_1}{x_2 - x_1}$ or

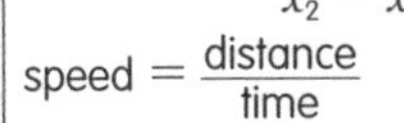

speed $= \frac{\text{distance}}{\text{time}}$

c What was the speed of the lift between 50 and 70 seconds?

21 ÷ .... = ... m/s

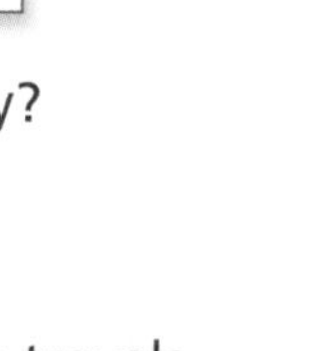

d Between what times was the fastest part of the journey?

Between .... and .... seconds

Practice

2 Steam trains run over a 15-mile section of track. Each train travels the 15 miles of track then makes the return journey to the station.

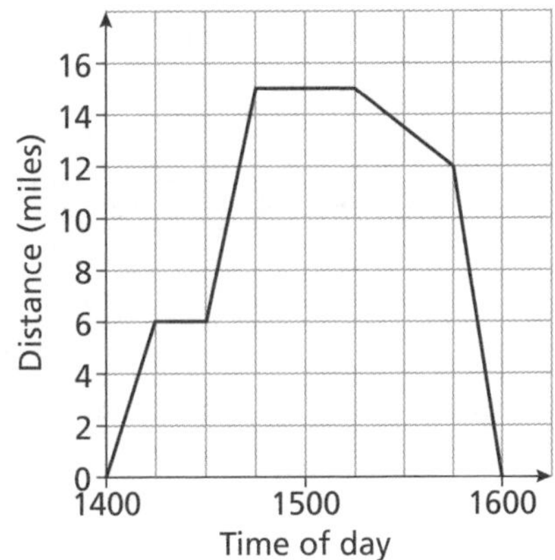

a The train first stopped at 1415 hours. For how long did it stop? .... minutes

b How far did the train travel between 1515 and 1600? .... miles

c On reaching the far end of the track the train lost water pressure and started the return journey slowly. What was the speed of the train over the slower section of its return journey?

.... mph

d To make up for lost time, the train then travelled at maximum speed. Work out this speed.

.... mph

3 The graph shows the first part of Tony's run.

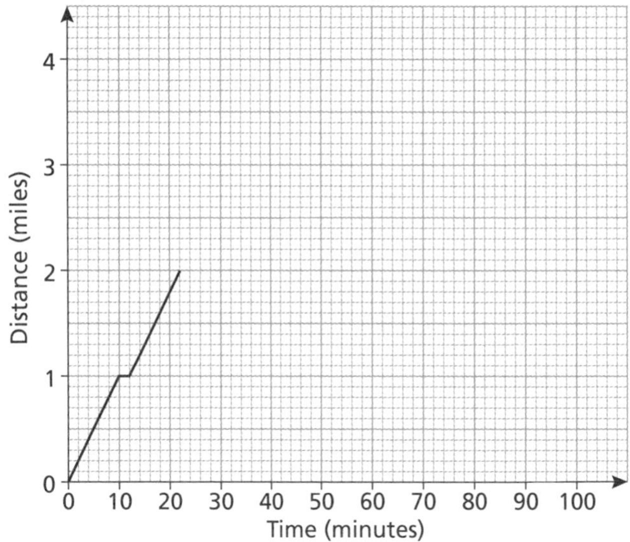

a Tony stopped after 10 minutes to rest. Estimate how long Tony stopped for.

.... minutes

b How long did Tony take to run the first mile?

.... minutes

c After Tony has run 2 miles he stops for a 4-minute stretch. He then runs a further 2 miles, which takes 22 minutes. After another 2-minute rest Tony runs home at a steady speed without stopping. It takes him 45 minutes. Complete the graph for the run.

4 Here are three graphs which show different parts of a car journey.

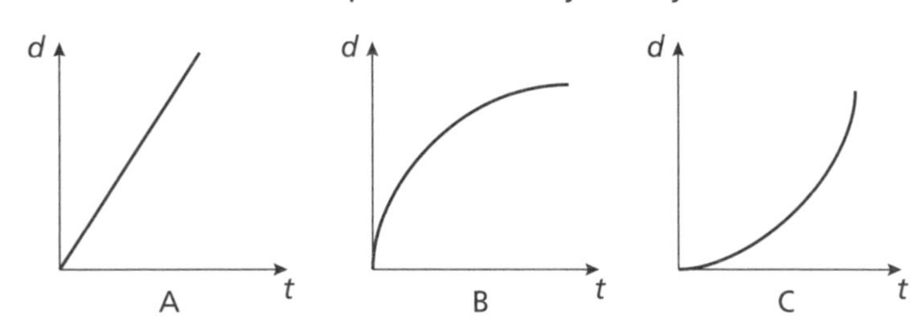

a Which graph shows the car travelling at an increasing speed? ....

b Which graph shows the car travelling at a constant speed? ....

Needs more practice ☐ Almost there ☐ I'm proficient! ☐

11.2 Speed–time graphs

By the end of this section you will know how to:

* Draw, interpret and understand speed–time graphs

AS LINKS

M1: 2.4 Representing the motion of an object on a speed–time or distance–time graph

Key points

* On a speed–time graph the y-axis represents the speed.
* On a speed–time graph the x-axis represents the time taken to travel.
* On a speed–time graph horizontal sections represent constant speed.
* The gradient of a line on a speed–time graph represents the acceleration.
* A positive gradient of a line on a speed–time graph represents acceleration.
* A negative gradient of a line on a speed–time graph represents deceleration.
* The area under a speed–time graph represents the distance travelled.

Guided

1 The speed–time graph represents the first stage of a cycle journey.

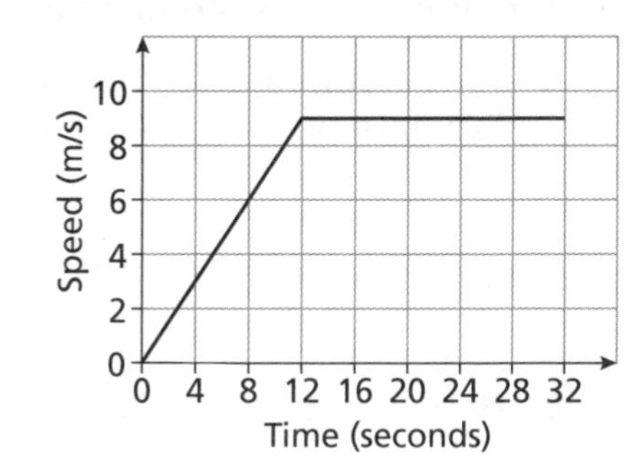

a Find the acceleration during the first part of the journey.

Acceleration = speed ÷ time

= 9 ÷ ....

= .... m/s²

b Find the distance travelled in the first 32 seconds.

Method 1

Distance = area of trapezium

$= \frac{1}{2} \times 9 \times ($ .... + ... $)$

= .... m

= .... m

Hint

Area of a trapezium $= \frac{1}{2}h(a + b)$, where h is the height of the trapezium and a and b are the lengths of the two parallel sides.

Method 2

Distance = area of triangle + area of ....

$= \frac{1}{2} \times 9 \times$ .... + 9 × ....

= .... + ...

= .... m

2 The graph represents a car journey in congested traffic.

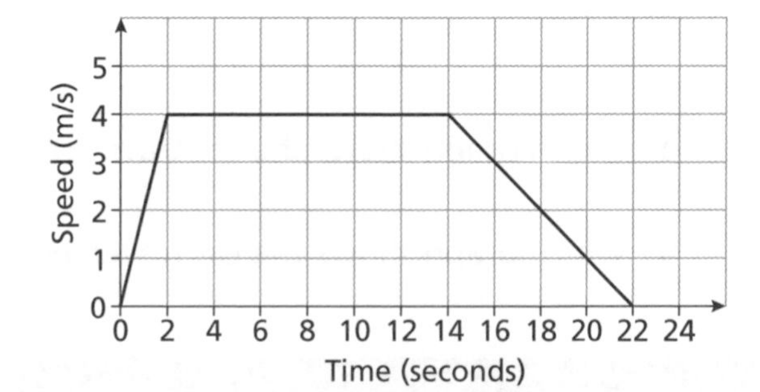

a What is the car's deceleration during the last 8 seconds of the journey?

Deceleration = ... ÷ 8

= .... m/s²

b What is the total distance covered by the car on this journey?

Distance = area of trapezium

$= \frac{1}{2} \times$ .... (.......... + ...)

= .... m

Practice

3 Part of a car journey is represented by this speed–time graph.

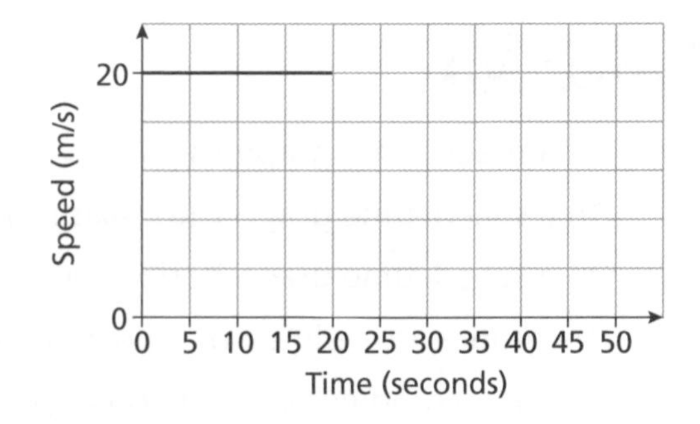

a After 20 seconds the car has to stop at a constant rate of deceleration which takes 10 seconds. Show this information on the graph.

b Calculate the car's deceleration during braking.

.... m/s²

c What distance does the car travel during the 10 seconds it is decelerating?

.... m

4 Part of a train journey is represented by this speed–time graph.

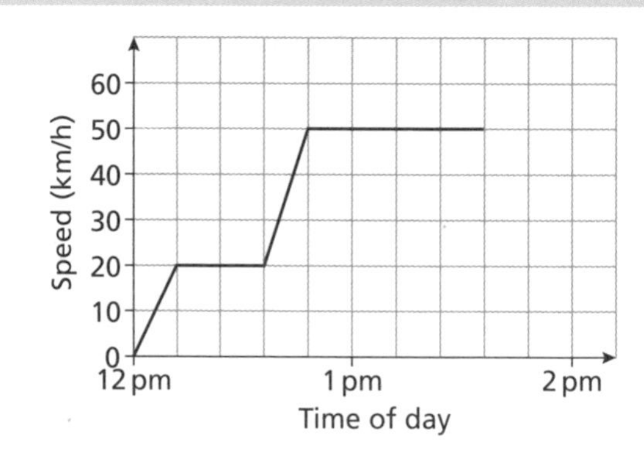

a What is the acceleration of the train for the first 12 minutes of the journey?

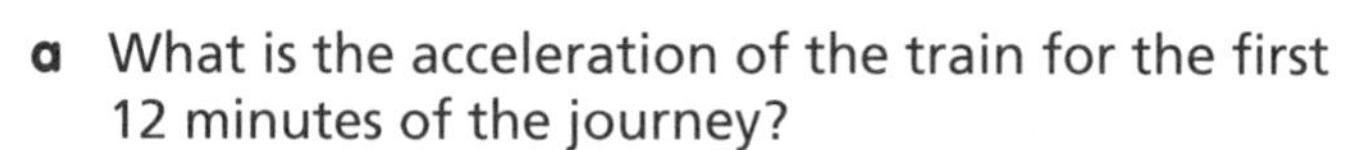

.... km/h^2

b What is the total distance covered during the journey between 12 pm and 1.36 pm?

.... km

5 A tractor journey is represented by this speed–time graph.

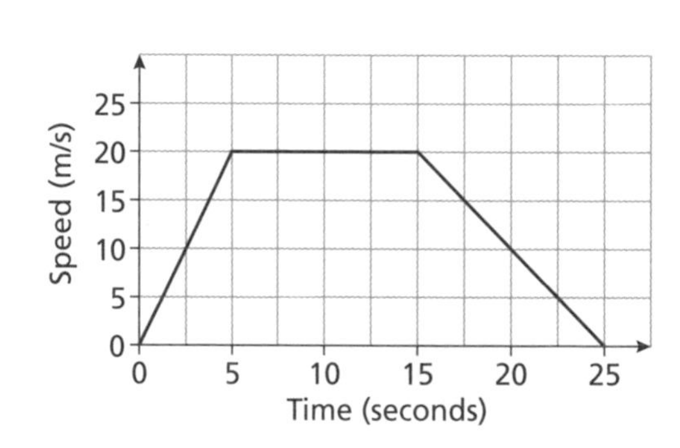

a Find the acceleration for the first 5 seconds.

.... m/s^2

b Find the distance travelled in the first 10 seconds.

.... m

c Find the deceleration at the time $t = 20$.

.... m/s^2

6 A cycle journey is represented by this speed–time graph.

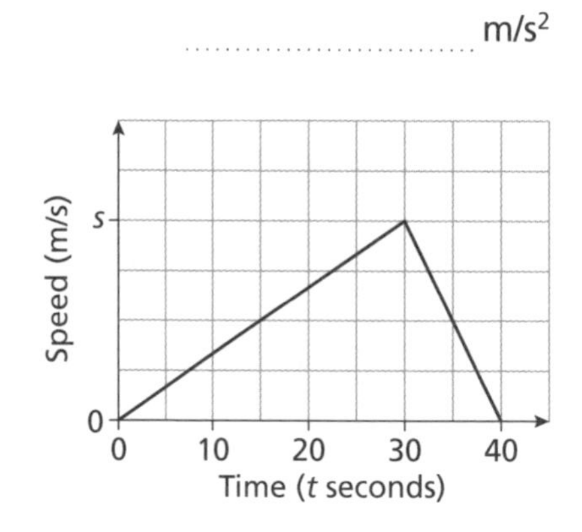

a The total distance travelled is 600 metres. Find the speed s at $t = 30$.

.... m/s

b What is the acceleration at $t = 10$?

.... m/s^2

Step into AS

7 The speed–time graph below is not drawn accurately.

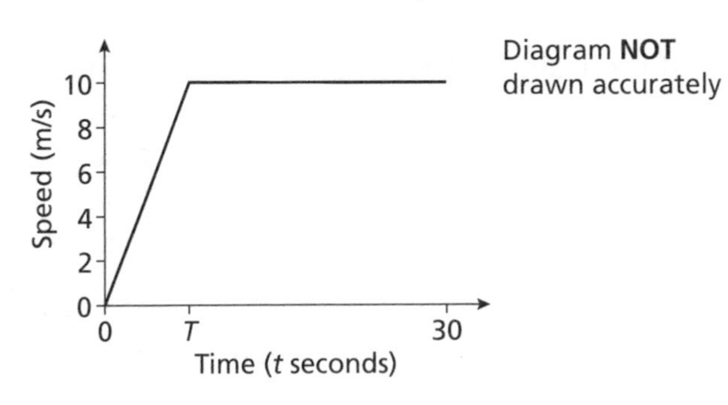

a The acceleration between 0 and T seconds is 2 m/s^2. Find the value of T.

.... s

b Find the total distance travelled between 0 and 30 seconds.

.... m

Don't forget!

- On a distance–time graph .... is plotted on the vertical axis.
- On a distance–time graph .... is plotted on the horizontal axis.
- On a distance–time graph the gradient of a line represents ....
- The steeper the line the .... the speed.
- On a distance–time graph horizontal sections represent ....
- On a speed–time graph .... is plotted on the vertical axis.
- On a speed–time graph .... is plotted on the horizontal axis.
- On a speed–time graph horizontal sections represent ....
- A positive gradient of a line on a speed–time graph represents ....
- A negative gradient of a line on a speed–time graph represents ....
- The area under a speed–time graph represents ....

Exam-style questions

1 This speed–time graph shows Anne's speed in her car, between her house and the first road junction.

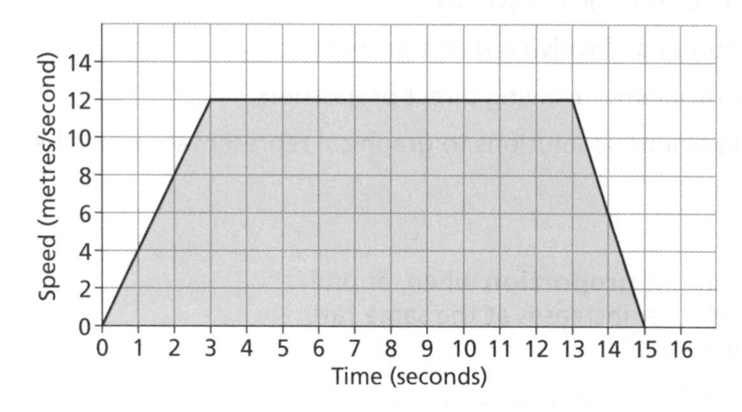

a Work out Anne's acceleration in the first 3 seconds.

.... metres per second^2

b Work out the total distance from Anne's house to the junction.

.... m

Needs more practice ☐ Almost there ☐ I'm proficient! ☐

12.1 Direct proportion

By the end of this section you will know how to:

- Find formulae involving direct proportion
- Solve problems involving direct proportion
- Relate algebraic solutions to graphical representations of the equations

Key points

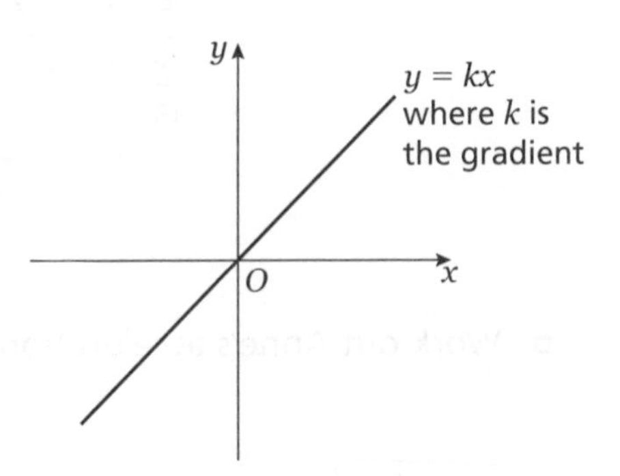

- Two quantities are in **direct proportion** when, as one quantity increases, the other increases at the same rate. Their **ratio** remains the same.
- 'y is directly proportional to x' is written as $y \propto x$. If $y \propto x$ then $y = kx$, where k is a constant.
- When x is directly proportional to y the graph is a straight line passing through the origin.

Guided

1 Paul gets paid at an hourly rate. The amount of pay (£P) is directly proportional to the number of hours (h) he works. When he works 8 hours he is paid £56.

a Find a formula to calculate Paul's pay.

> **Hint**
> Substitute the values given for P and h into the formula to calculate k.

$P \propto$ ....

$P = k$..........

$56 = k \times$ ....

$k = 56 \div$ .... $=$ ....

So the formula is $P =$ .... h

b If Paul works for 11 hours, how much is he paid?

$P =$ .... h

$P =$ .... $\times$ ....

$P =$ £..........

2 y is directly proportional to x^2. When $x = 3$, $y = 45$.

a Find a formula for y in terms of x.

$y \propto x^2$

$y =$ .... x^2

.... $=$ .... $\times$ ....^{2}

$k =$ .... $\div$ .... $=$ ....

So the formula is $y =$ .... x^2

b Find y when $x = 5$.

$y =$ .... x^2

$y =$ .... $\times$ ....^{2}

$y =$ ....

c Find x when $y = 20$.

$y =$ .... x^2

.... $=$ .... $\times x^2$

$x^2 =$ .... $\div$ .... $=$ ....

$x =$ ....

Practice

3 x is directly proportional to y.

a Find a formula for x in terms of y. x is 35 when y is 5.

b Sketch the graph of the formula.

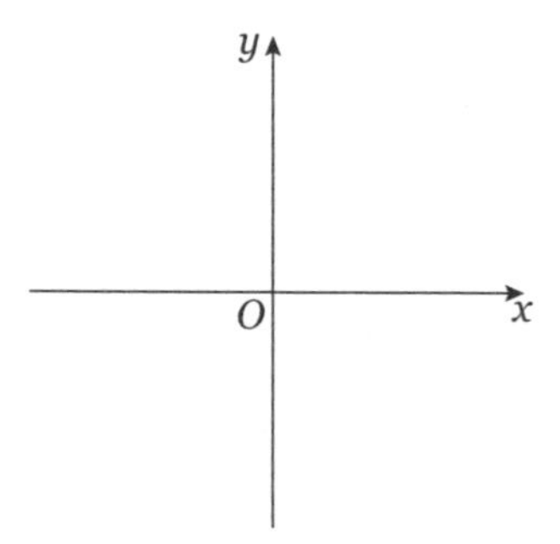

c Find x when y is 13.

d Find y when x is 63.

4 Q is directly proportional to the square of Z. $Q = 48$ when $Z = 4$.

a Find a formula for Q in terms of Z.

b Sketch the graph of the formula.

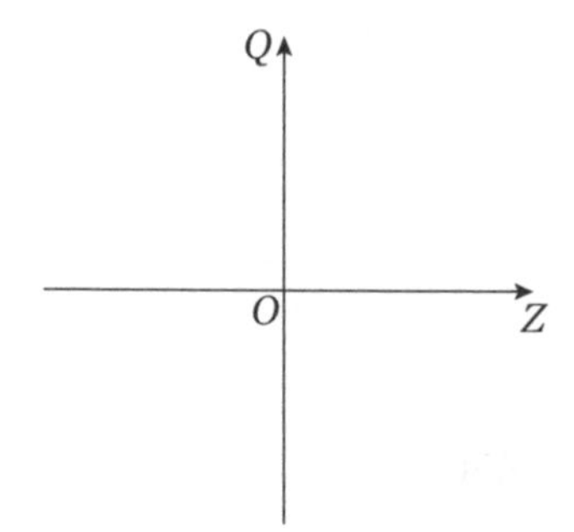

c Find Q when $Z = 5$.

d Find Z when $Q = 300$.

5 y is directly proportional to the square of x. x is 2 when y is 10.

a Find a formula for y in terms of x.

b Sketch the graph of the formula.

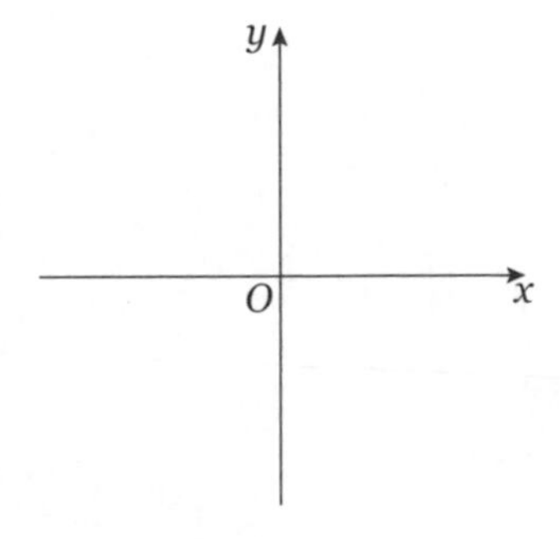

c Find x when y is 90.

6 B is directly proportional to the square root of C. $C = 25$ when $B = 10$.

a Find a formula for B in terms of C.

b Find B when $C = 64$.

c Find C when $B = 20$.

7 C is directly proportional to D. $C = 100$ when $D = 150$.

a Find a formula for C in terms of D.

b Find C when $D = 450$.

8 y is directly proportional to x. $x = 27$ when $y = 9$.

a Find a formula for x in terms of y.

b Find x when $y = 3.7$.

9 m is proportional to the cube of n. $m = 54$ when $n = 3$.

a Find a formula for m in terms of n.

b Find n when $m = 250$.

Needs more practice ☐ Almost there ☐ I'm proficient! ☐

12.2 Inverse proportion

By the end of this section you will know how to:

- Find formulae involving inverse proportion
- Solve problems involving inverse proportion
- Relate algebraic solutions to graphical representations of the equations

Key points

- Two quantities are in **inverse proportion** when, as one quantity increases, the other decreases at the same rate.
- 'y is inversely proportional to x' is written as $y \propto \frac{1}{x}$. If $y \propto \frac{1}{x}$ then $y = \frac{k}{x}$, where k is a constant.
- When x is inversely proportional to y the graph is the same shape as the graph of $y = \frac{1}{x}$.

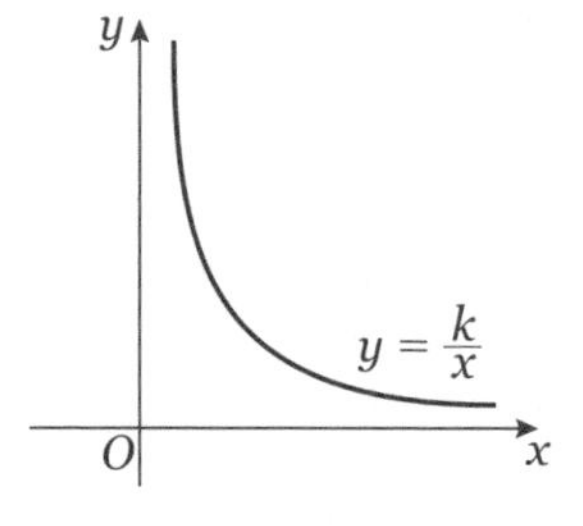

Guided

1 P is inversely proportional to Q. $P = 100$ when $Q = 10$.

a Find a formula for P in terms of Q.

$P \propto \frac{1}{Q}$

$P = \frac{k}{Q}$

.... $= \frac{k}{......}$

$k =$ .... $\times$ ....

$k =$ ....

$P = \frac{..........}{Q}$

Hint

Substitute the values given for and into the equation to calculate .

b Find Q when $P = 20$.

$P = \frac{..........}{Q}$

.... $= \frac{..........}{Q}$

$Q =$ .... $\div$ ....

$Q =$ ....

2 y is inversely proportional to the square root of x.
When $y = 1$, $x = 25$.

a Find a formula for y in terms of x.

$y \propto \frac{1}{\sqrt{\ldots\ldots}}$

$y = \frac{\ldots\ldots}{\sqrt{\ldots\ldots}}$

$\ldots\ldots\ldots = \frac{\ldots\ldots}{\sqrt{25}}$

$k = \ldots\ldots\ldots \times \ldots\ldots\ldots$

$k = \ldots\ldots\ldots$

$y = \frac{\ldots\ldots}{\sqrt{x}}$

b Find x when $y = 5$.

$y = \frac{\ldots\ldots}{\sqrt{x}}$

$\ldots\ldots\ldots = \frac{\ldots\ldots}{\sqrt{x}}$

$\sqrt{x} = \ldots\ldots\ldots \div \ldots\ldots\ldots = \ldots\ldots\ldots$

$x = \ldots\ldots\ldots$

Practice

3 s is inversely proportional to t.

a Given that $s = 2$ when $t = 2$, find a formula for s in terms of t.

b Sketch the graph of the formula.

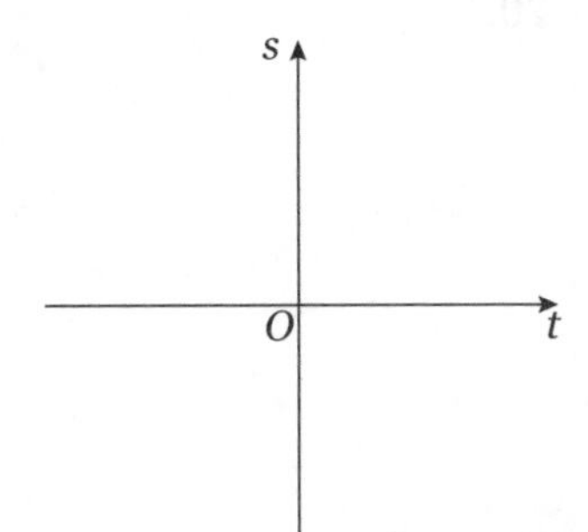

c Find t when $s = 1$.

4 a is inversely proportional to b.

a Given that $a = 5$ when $b = 20$, find a formula for a in terms of b.

b Find a when $b = 50$.

c Find b when $a = 10$.

5 v is inversely proportional to w. $w = 4$ when $v = 20$.

a Find a formula for v in terms of w.

b Sketch the graph of the formula.

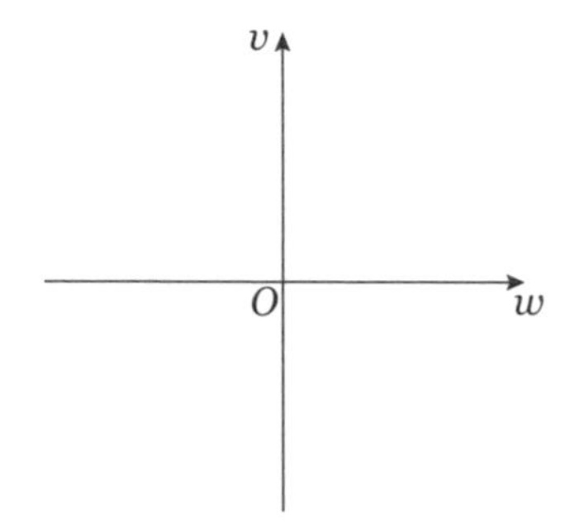

c Find w when $v = 2$.

6 L is inversely proportional to W. $L = 12$ when $W = 3$.

a Find a formula for L in terms of W.

b Find W when $L = 6$.

7 s is inversely proportional to t.

a Given that $s = 6$ when $t = 12$, find a formula for s in terms of t.

b Find s when $t = 3$.

c Find t when $s = 18$.

8 y is inversely proportional to x^2. $y = 4$ when $x = 2$.

a Find a formula for y in terms of x.

b Find y when $x = 4$.

9 a is inversely proportional to b. $a = 0.05$ when $b = 4$.

a Find a formula for a in terms of b.

b Find a when $b = 2$.

c Find b when $a = 2$.

Don't forget!

* Two quantities are in .... proportion when, as one quantity increases, the other increases at the same rate.
* Two quantities are in .... proportion when, as one quantity increases, the other decreases at the same rate.
* The sign used for proportion is ....
* 'y is directly proportional to x' is written as y .... If $y \propto x$ then y ...., where k is a constant.
* 'y is inversely proportional to x' is written as y .... If $y \propto \frac{1}{x}$ then y ...., where k is a constant.
* When x is directly proportional to y the graph looks like:

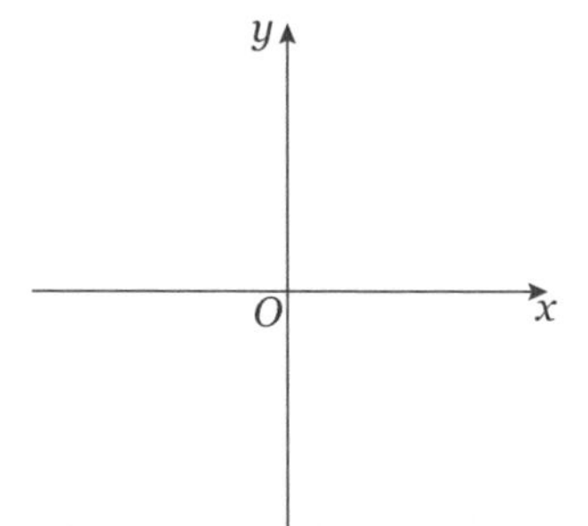

* When x is inversely proportional to y the graph looks like:

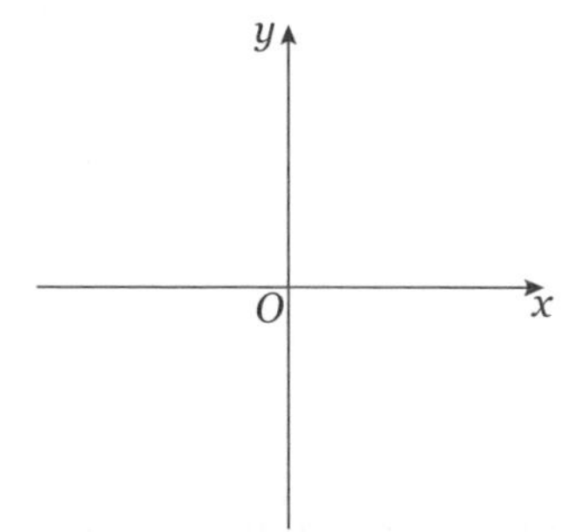

Exam-style questions

1 A is directly proportional to the square of B.
When $A = 48$, $B = 4$

a Find a formula for A in terms of B.

$A =$ ....

b Calculate the value of A when $B = \frac{1}{2}$

....

c Calculate the value of B when $A = 1.08$

....

Needs more practice ☐ Almost there ☐ I'm proficient! ☐

13.1 Applying the transformations $y = f(x) \pm a$ and $y = f(x \pm a)$ to the graph of $y = f(x)$

AS LINKS

C1: 4.5 The effect of the transformations $y = f(x) + a$ and $y = f(x \pm a)$; 4.7 Performing transformations on the sketches of curves

By the end of this section you will know how to:

* Apply the transformations $y = f(x) \pm a$ and $y = f(x \pm a)$ to the graph of $y = f(x)$

Key points

* The transformation $y = f(x) \pm a$ is a translation of $y = f(x)$ parallel to the y-axis; it is a vertical translation. As shown on the graph below, $y = f(x) + a$ translates $y = f(x)$ up and $y = f(x) - a$ translates $y = f(x)$ down.

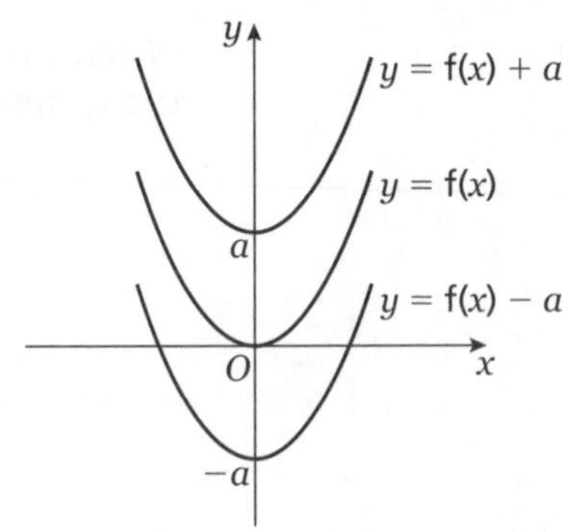

* The transformation $y = f(x \pm a)$ is a translation of $y = f(x)$ parallel to the x-axis; it is a horizontal translation. As shown on the graph below, $y = f(x + a)$ translates $y = f(x)$ to the left and $y = f(x - a)$ translates $y = f(x)$ to the right.

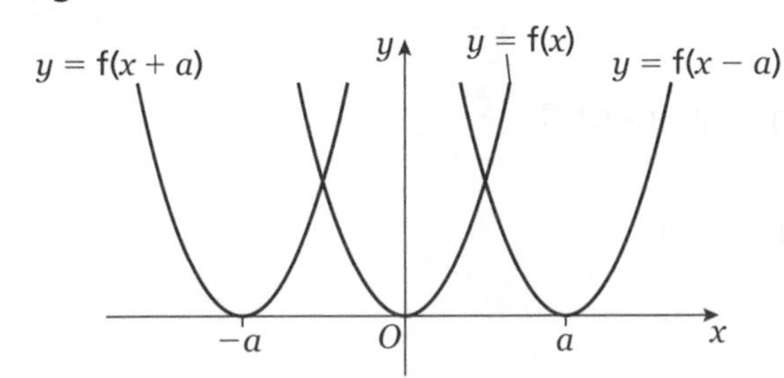

Guided

1 The graph shows the function= f(). Sketch the graph of= f() + 2.

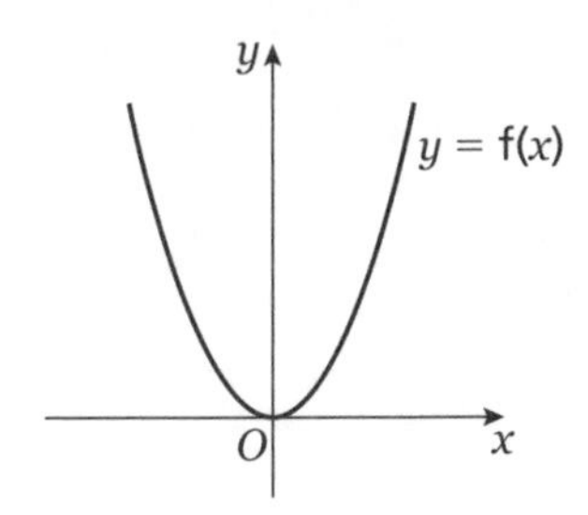

Hint

For the function $y = f(x) + 2$ translate the curve 2 units up.

2 The graph shows the function= f(). Sketch the graph of= f(− 3).

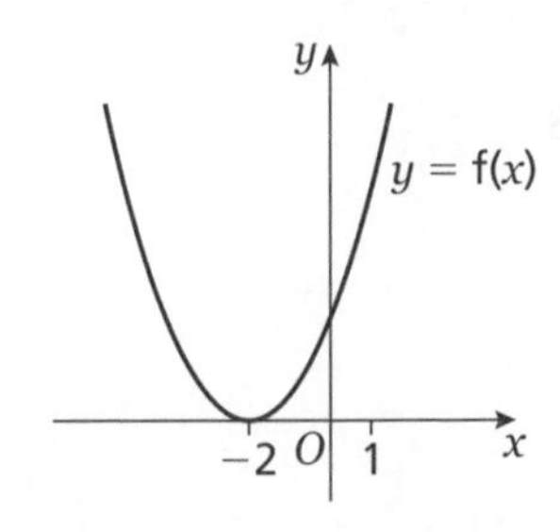

Hint

For the function $y = f(x - 3)$ translate the curve 3 units right.

Practice

3 The graph shows the function= f().
On the same axes, sketch and label the graphs of= f() + 4 and= f(+ 2).

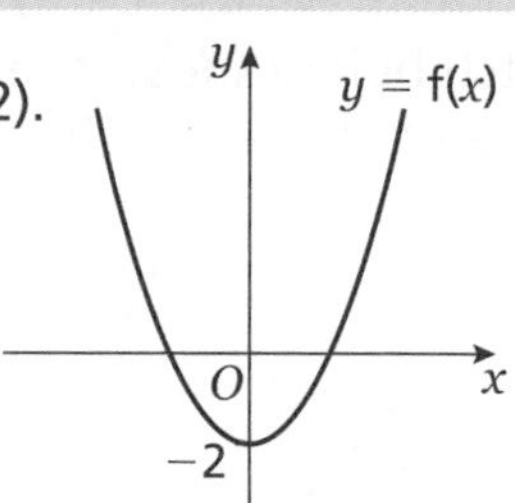

4 The graph shows the function= f().
On the same axes, sketch and label the graphs of= f(+ 3) and= f() − 3.

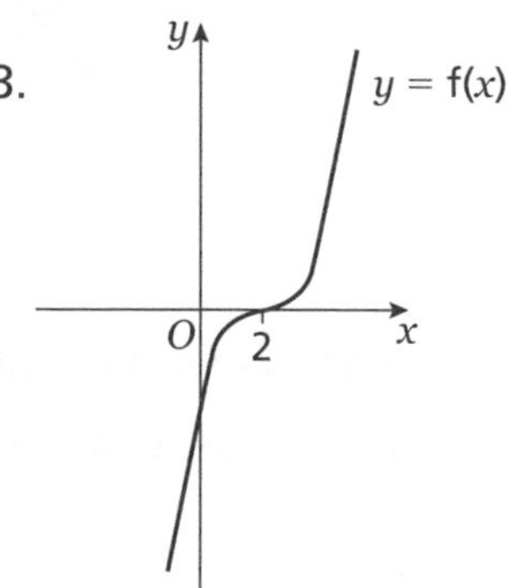

5 The graph shows the function= f().
Sketch the graph of= f(− 5).

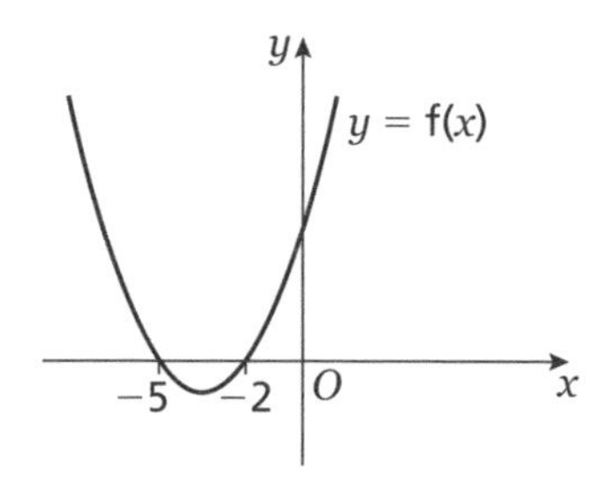

6 The sketch below shows the function= f() and two transformations of= f(), labelled $_1$ and $_2$. Write down the equations of the translated curves and $_2$ in function form.

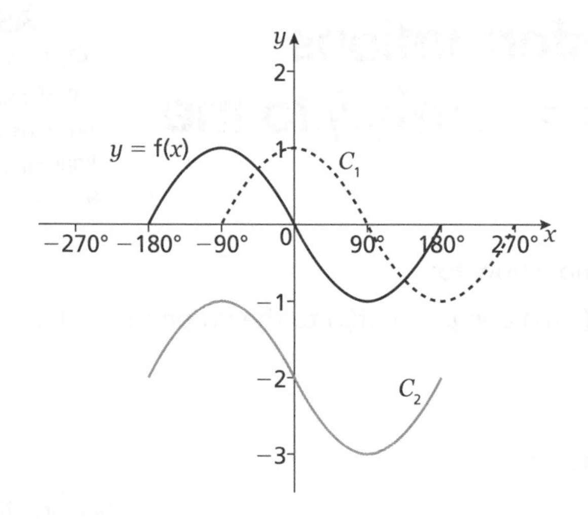

7 The sketch below shows the function= f() and two transformations of= f(), labelled $_1$ and $_2$. Write down the equations of the translated curves $_1$ and $_2$ in function form.

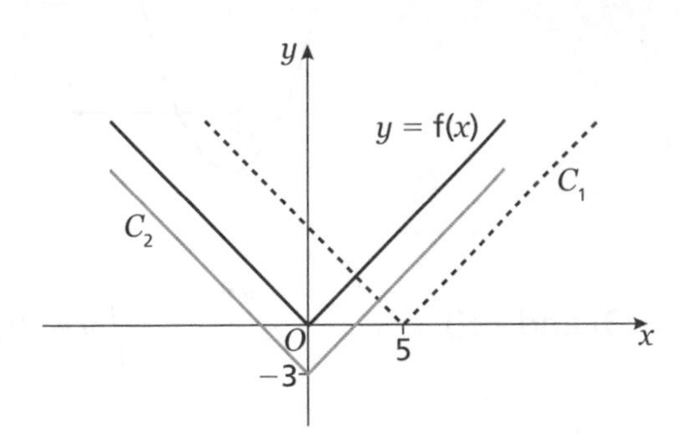

8 **a** Sketch and label the graph of= f(), where f() = (− 1)(+ 1).

b On the same axes, sketch and label the graphs of= f() − 2 and= f(+ 2).

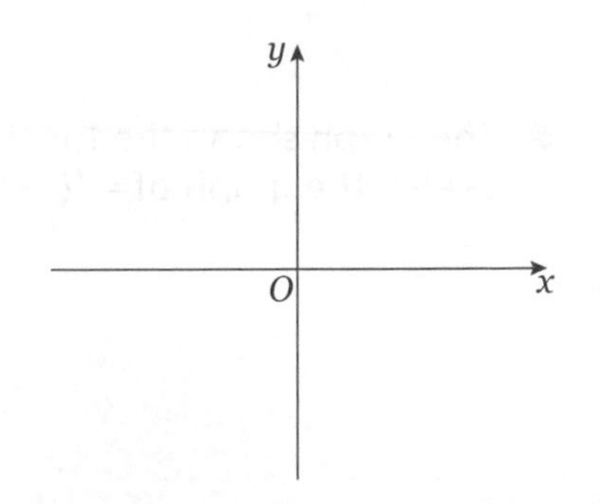

Needs more practice ☐ Almost there ☐ I'm proficient! ☐

13.2 Applying the transformations $y = f(\pm ax)$ and $y = \pm af(x)$ to the graph of $y = f(x)$

AS LINKS
C1: 4.6 The effect of the transformations f(ax) and af(x); 4.7 Performing transformations on the sketches of curves

By the end of this section you will know how to:

* Apply the transformations $y = f(\pm ax)$ and $y = \pm af(x)$ to the graph of $y = f(x)$

Key points

* The transformation $y = f(ax)$ is a horizontal stretch of $y = f(x)$ with scale factor $\frac{1}{a}$ parallel to the x-axis.

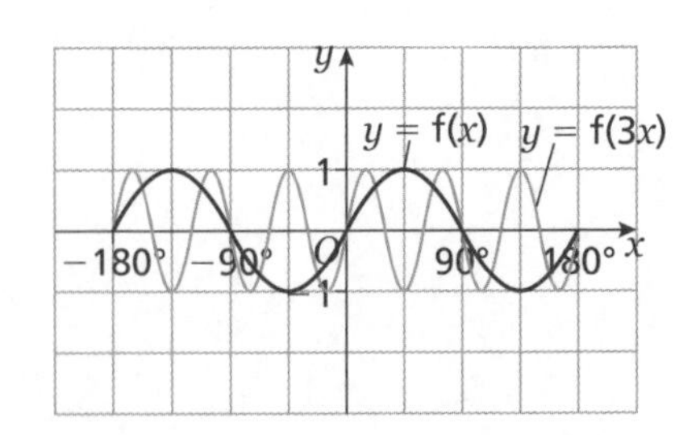

* The transformation $y = f(-ax)$ is a horizontal stretch of $y = f(x)$ with scale factor $\frac{1}{a}$ parallel to the x-axis and then a reflection in the y-axis.

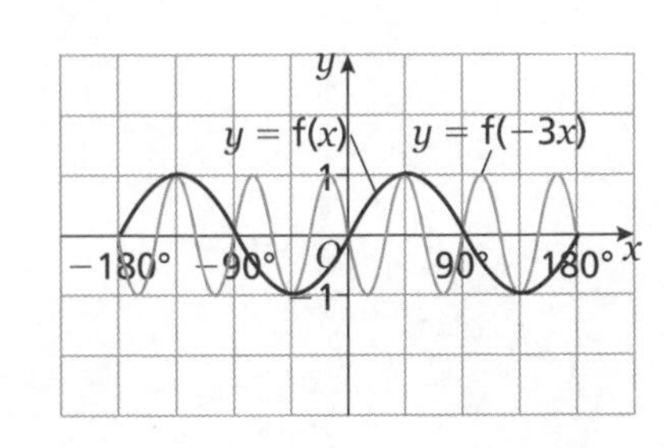

★ The transformation $y = af(x)$ is a vertical stretch of $y = f(x)$ with scale factor a parallel to the y-axis.

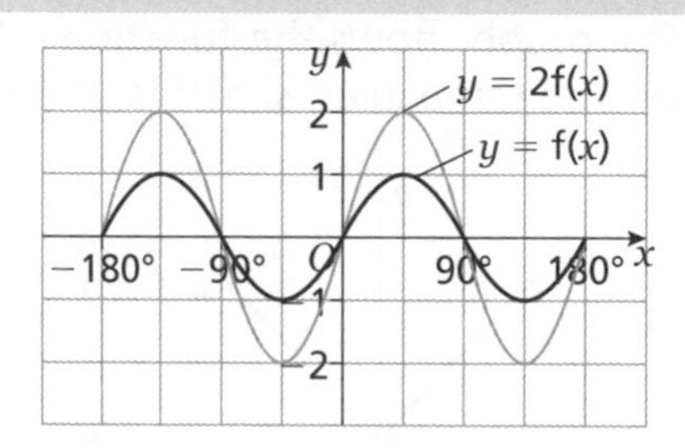

★ The transformation $y = -af(x)$ is a vertical stretch of $y = f(x)$ with scale factor a parallel to the y-axis and then a reflection in the x-axis.

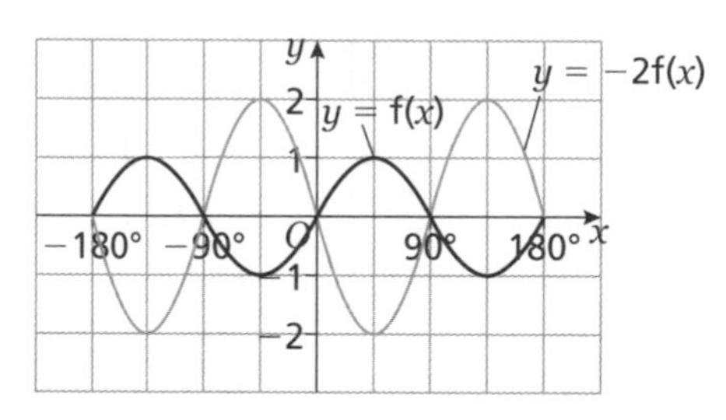

Guided

1 The graph shows the function= f().
On the same axes, sketch and label the graphs of= 2f() and= −f().

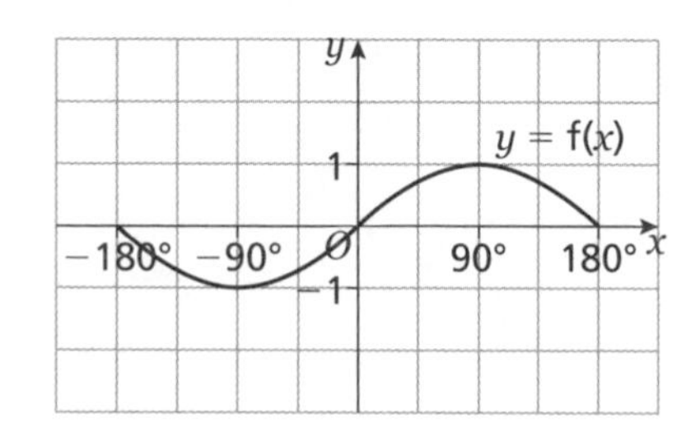

Hint
The function $y = -f(x)$ is a reflection in the x axis of $y = f(x)$.

2 The graph shows the function= f().
On the same axes, sketch and label the graphs of= f(2) and= f(−).

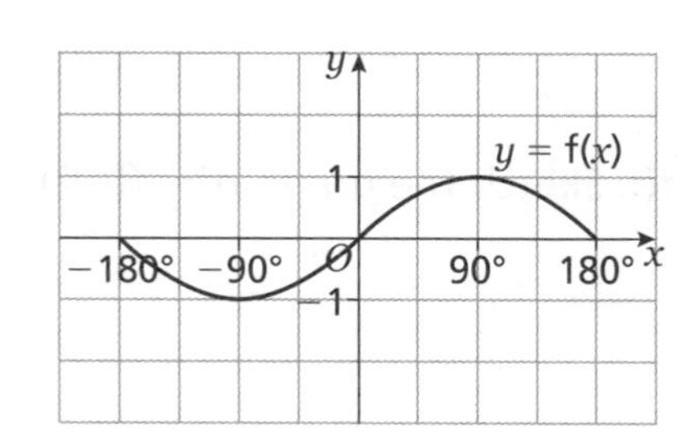

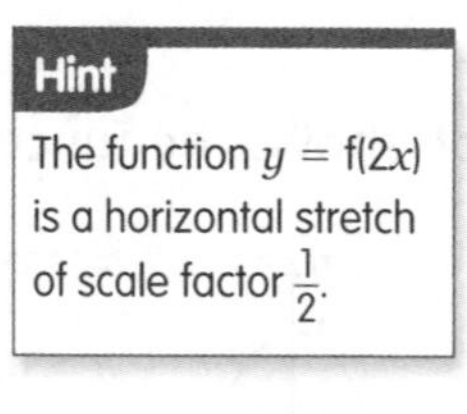
Hint
The function $y = f(2x)$ is a horizontal stretch of scale factor $\frac{1}{2}$.

Practice

3 The graphs show the function= f().

a Sketch and label the graph of= 3f().

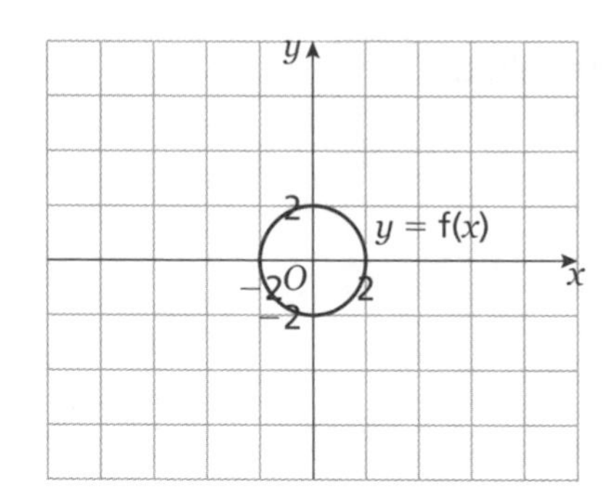

b Sketch and label the graph of= f(2).

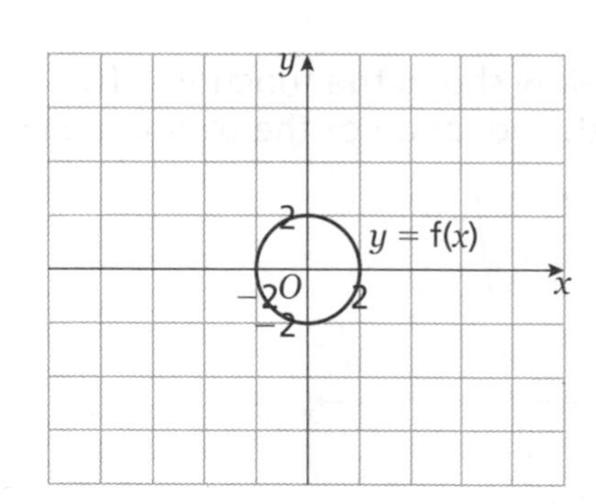

4 The graph shows the function= f().
On the same axes, sketch and label the graphs of= −2f() and= f(3).

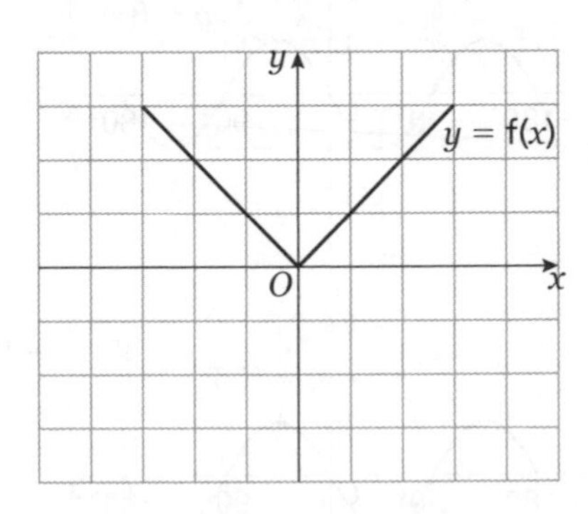

5 The graph shows the function= f().
On the same axes, sketch and label the graphs of= −f() and= f($\frac{1}{2}$).

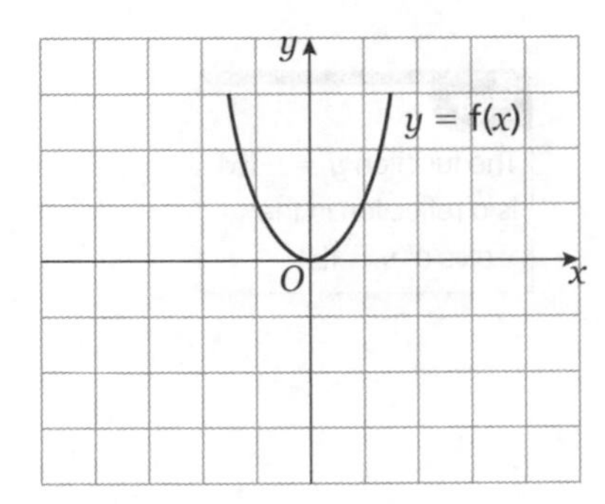

6 The graph shows the function= f(). Sketch the graph of= f(−2).

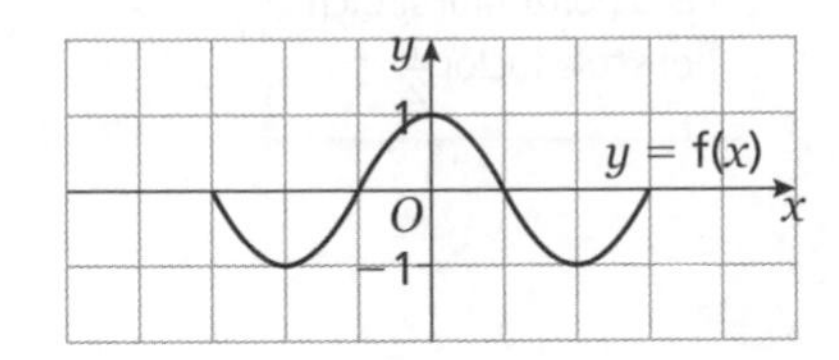

7 The sketch below shows the function= f() and a transformation, labelled .
Write down the equation of the translated curve in function form.

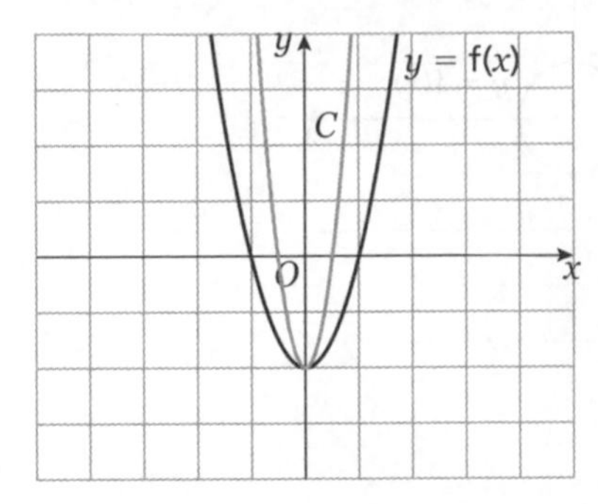

8 The sketch below shows the function= f() and a transformation, labelled .
Write down the equation of the translated curve in function form.

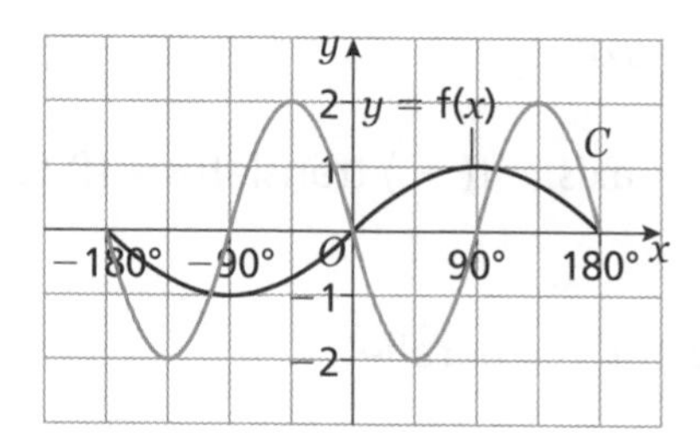

9 a Sketch and label the graph of= f(), where f() = cos **b** On the same axes, sketch and label the graph of= −2f().

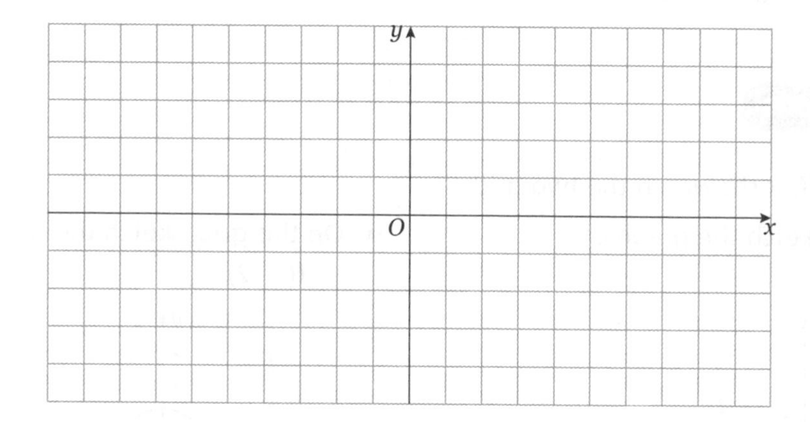

10 a Sketch and label the graph of= f(), where f() = −(+ 1)(− 2)

b On the same axes, sketch and label the graph of= f(−$\frac{1}{2}$).

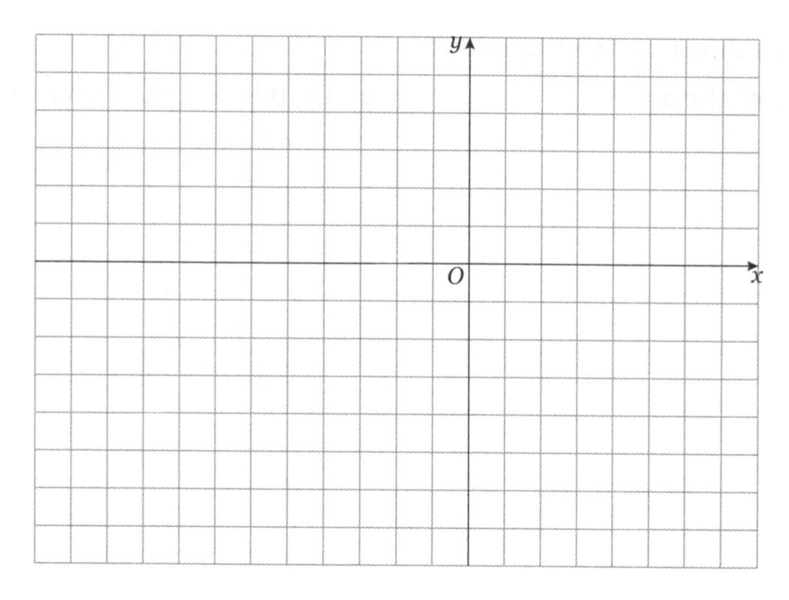

Don't forget!

- The transformation = f() ± is a translation of= f() parallel to the ...... -axis.
- The transformation = f(±) is a translation of= f() parallel to the ...... -axis.= f(+) translates= f() to the ...... and= f(− translates= f() to the ......
- The transformation = f() is a horizontal stretch of= f() with scale factor ...... parallel to the ...... -axis.
- The transformation = f(−) is a horizontal stretch of= f() with scale factor ...... parallel to the ...... -axis and then a reflection in the ...... axis.
- The transformation = f() is a vertical stretch of= f() with scale factor ...... parallel to the ...... axis.
- The transformation = −f() is a vertical stretch of= f() with scale factor ...... parallel to the ...... axis and then a reflection in the ...... axis.

Exam-style questions

1 The graph of = f() is shown on the two grids.

a On this grid, sketch the graph of = f() + 2

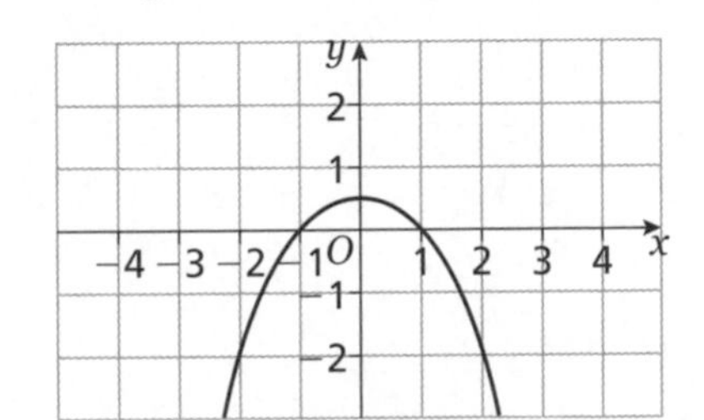

b On this grid, sketch the graph of = f(+ 2)

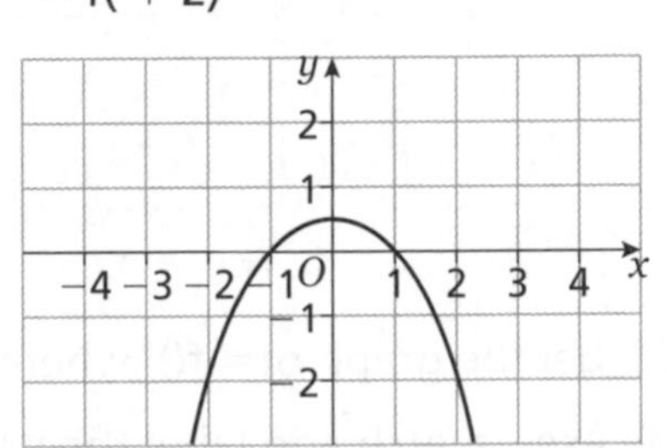

2 The graph of = f() is shown on the two grids.

a On this grid, sketch the graph of = −f()

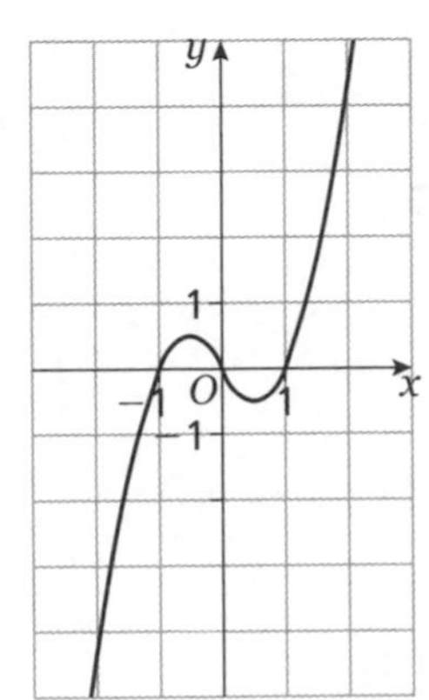

b On this grid, sketch the graph of = 2f()

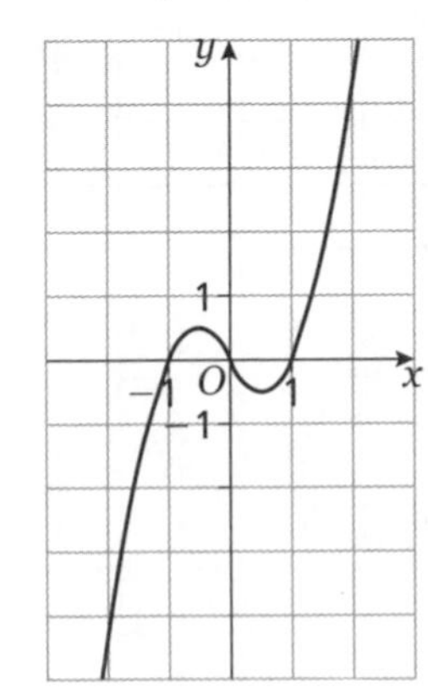

14.1 The trapezium rule

AS LINKS
C2: 11.5 The trapezium rule

By the end of this section you will know how to:

- Find an approximation for the area under a curve using the trapezium rule

Key points

- Using the **trapezium rule** gives us an approximation to the area under a curve.
- The trapezium rule is: Area $= \frac{1}{2}h[y_0 + 2(y_1 + y_2 \dots + y_{n-1}) + y_n]$
  where h is the width of each strip and $y_0, y_1, y_2 \dots y_{n-1}, y_n$ are the values of y for each value of x used.

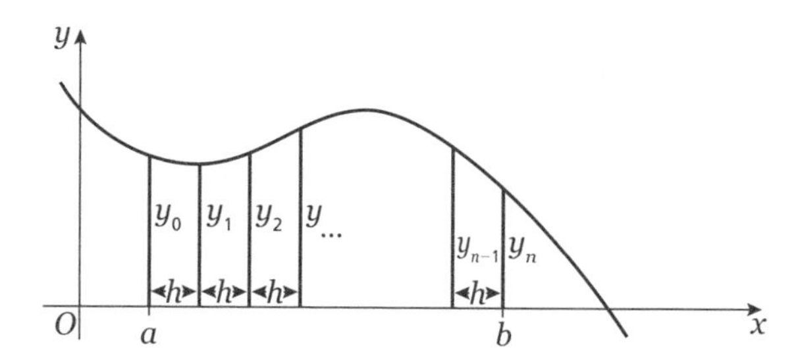

- n is the number of equal strips the area has been divided up into; $x = a$ and $x = b$ define the vertical boundaries of the area.
- The number of values used for x and the number of values used for y will be 1 more than the number of strips, n.
- The width of each strip, h, can be calculated using $h = \frac{b - a}{n}$

Guided

1 Use the trapezium rule to estimate the area of the region between the curve $y = (3 - x)(2 + x)$ and the x-axis from $x = 0$ to $x = 3$. Use 3 strips of equal width.

Each strip will be of width $h = \frac{b - a}{n} = \frac{3 - 0}{3} = \dots\dots$

Use a table to work out y for each value of x.

x	0	1	2	3
$y = (3 - x)(2 + x)$	6	....	....	0

$A = \frac{1}{2}h[y_0 + 2(y_1 + y_2 \dots + y_{n-1}) + y_n]$

In this case $A = \frac{1}{2}h[y_0 + 2(y_1 + y_2) + y_3]$

From the table above, $y_0 = 6$, $y_1 = \dots\dots$, $y_2 = \dots\dots$, $y_3 = 0$.

Substituting these values into the formula gives

$A = \frac{\dots\dots}{2} \times \dots\dots \times [6 + 2(6 + \dots\dots) + 0]$

$= \frac{\dots\dots}{2} [\dots\dots]$

$= \dots\dots$ square units

Hint
For a full answer, remember to include 'square units'.

2 Use the trapezium rule to estimate the shaded area. Use 3 strips of equal width.

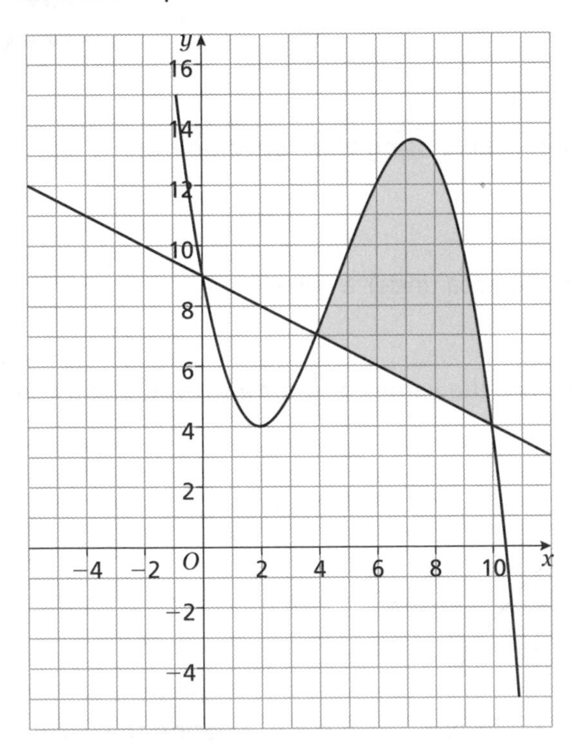

Each strip will be of width $h = \frac{b - a}{n} = \frac{10 - 4}{3} = \ldots\ldots$

Use a table to record y for each value of x.

x	4	6	8	10
y coordinate for the curve	7	....	....	4
y coordinate for the straight line	7	....	....	4

$A = \frac{1}{2} h[y_0 + 2(y_1 + y_2 \ldots + y_{n-1}) + y_n]$

In this case $A = \frac{1}{2} h[y_0 + 2(y_1 + y_2) + y_3]$

From the table above, $y_0 = 0$, $y_1 = \ldots\ldots$, $y_2 = \ldots\ldots$, $y_3 = 0$.

Substituting these values into the formula gives

Hint

Find the difference in the two y coordinates to find y_0 to y_3.

$A = \frac{\ldots\ldots}{2} \times \ldots\ldots [0 + 2(\ldots\ldots + \ldots\ldots) + 0]$

$= \ldots\ldots \times \ldots\ldots$

$= \ldots\ldots$ square units

Practice

3 Use the trapezium rule to estimate the area between the curve $y = (5 - x)(x + 2)$ and the x-axis from $x = 1$ to $x = 5$. Use 4 strips of equal width.

4 Use the trapezium rule to estimate the shaded area shown on the axes. Use 6 strips of equal width.

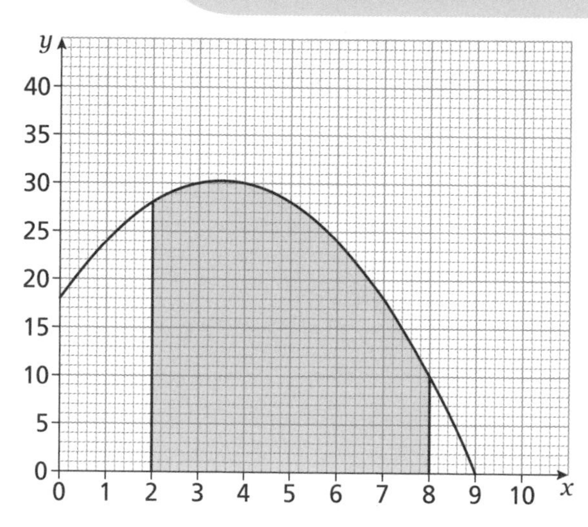

5 Use the trapezium rule to estimate the area between the curve $y = x^2 - 8x + 18$ and the x-axis from $x = 2$ to $x = 6$. Use 4 strips of equal width.

6 Use the trapezium rule to estimate the shaded area using 6 strips of equal width.

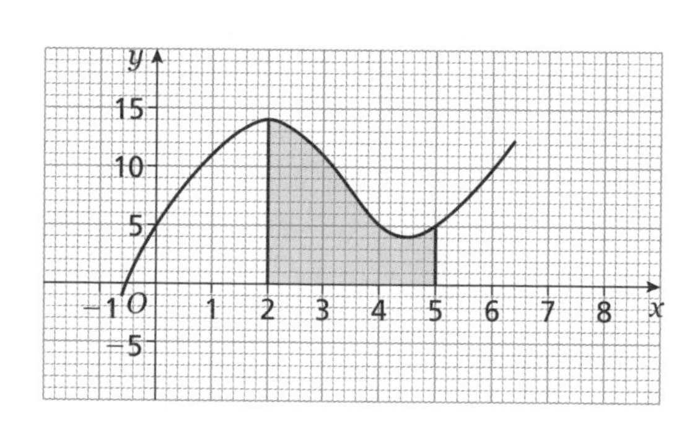

7 Use the trapezium rule to estimate the area between the curve $y = -x^2 - 4x + 5$ and the x-axis from $x = -5$ to $x = 1$. Use 6 strips of equal width.

8 Use the trapezium rule to estimate the shaded area using 4 strips of equal width.

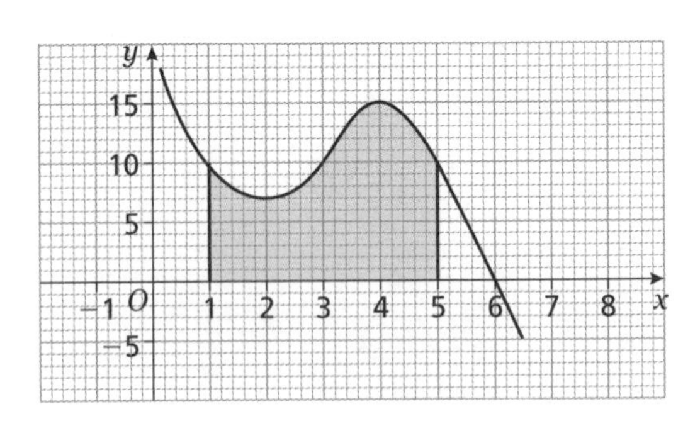

9 Use the trapezium rule to estimate the area between the curve $y = -x^2 + 2x + 15$ and the x-axis from $x = 2$ to $x = 5$. Use 6 strips of equal width.

10 Use the trapezium rule to estimate the shaded area using 7 strips of equal width.

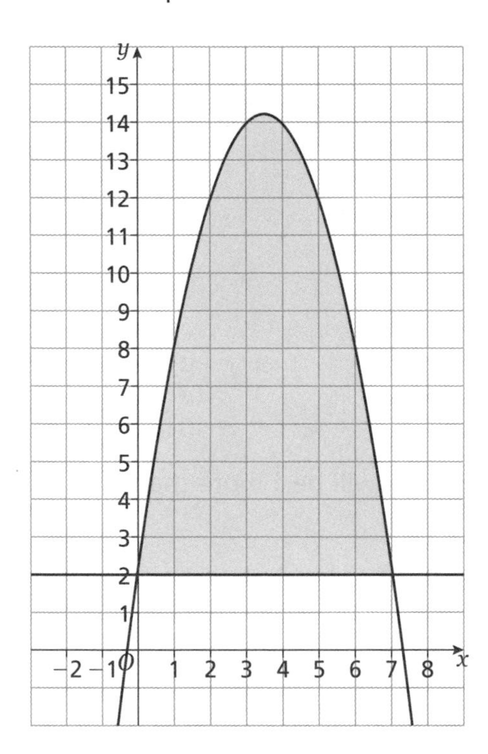

11 Use the trapezium rule to estimate the shaded area using 5 strips of equal width.

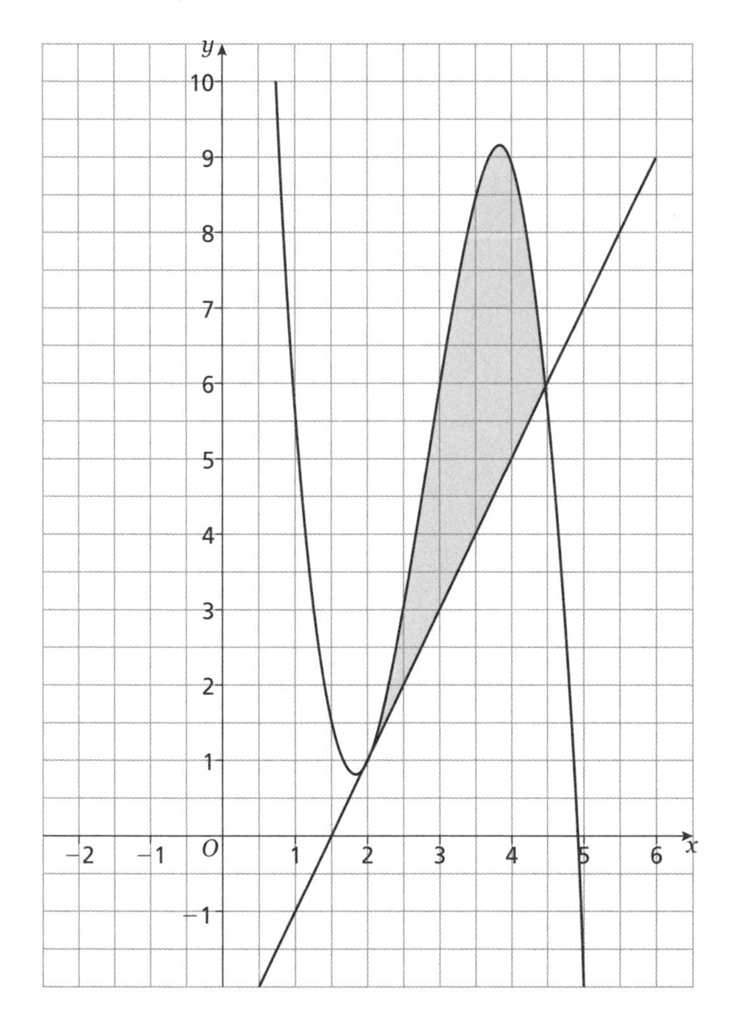

Don't forget!

* Using the trapezium rule gives us an approximation to ....
* The trapezium rule is: ....

  where h is the width of each strip and $y_0, y_1, y_2 \dots y_{n-1}, y_n$ are ....

  ....
* n is the .... ;

  $x = a$ and $x = b$ define ....
* The number of values used for x and the number of values used for y will be 1 more than

  ....
* The width of seach strip, h, can be calculatd using $h = \frac{\dots}{\dots}$

Exam-style questions

1 Use the trapezium rule to estimate the area of the shaded region.
Use 3 strips of equal width.

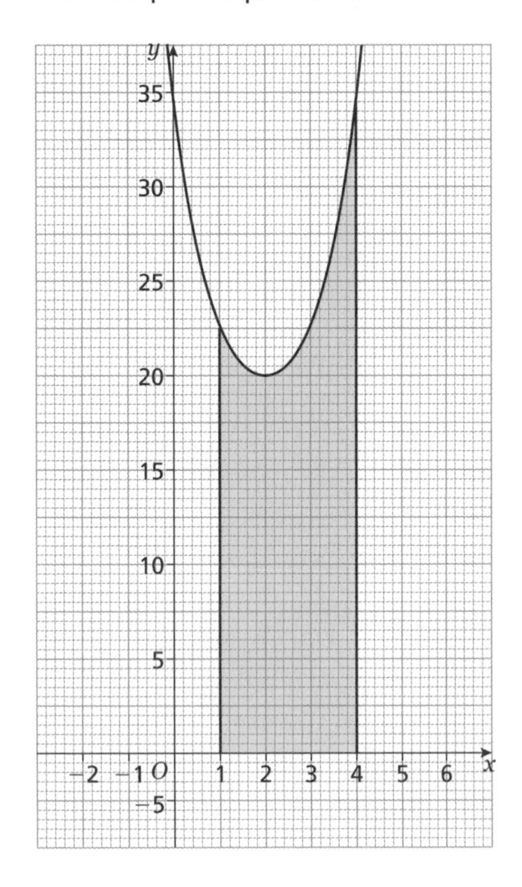

.... square units

2 The curve $y = 8x - 5 - x^2$ and the line $y = 2$ are shown in the sketch.

Use the trapezium rule with 6 strips to estimate the shaded area.

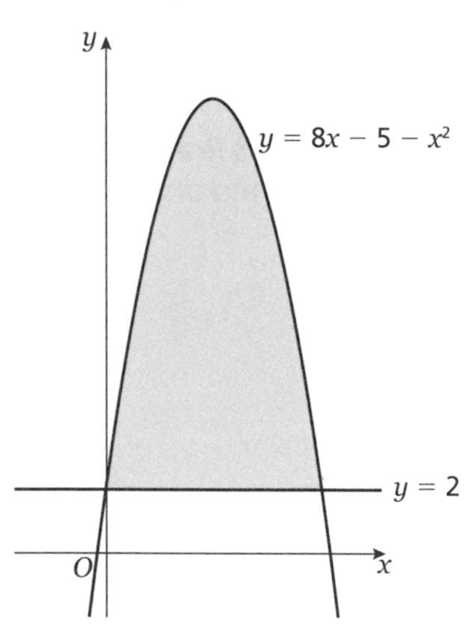

.... square units

3 Use the trapezium rule with 7 strips to find an estimate for the shaded area.

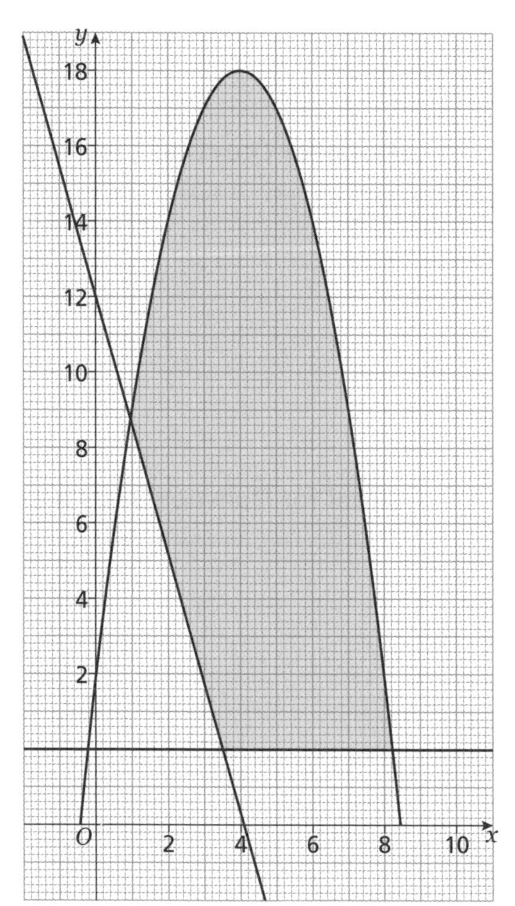

.... square units

Practice Paper

Time: 2 hours

Edexcel publishes Sample Assessment Material on its website. This Practice Exam Paper has been written to help you practise what you have learned and may not be representative of a real exam paper.

1 Solve the simultaneous equations

$2x - 3y = 7$

$y = 2x + 3$

....

(Total for Question 1 is 3 marks)

---

2 Make m the subject of the formula $k = 6m^2$

$m =$ ....

(Total for Question 2 is 2 marks)

---

3 Expand and simplify $4xy - (3x - y)(2x + 4y)$

....

(Total for Question 3 is 3 marks)

---

4 **a** Factorise $2x^2 - x - 3$

....

b Work out the value of $2 \times 16.5^2 - 16.5 - 3$

....

(Total for Question 4 is 3 marks)

---

5 Here is the graph of $y = f(x)$
The graph has a minimum point at (3, 0).

(a)

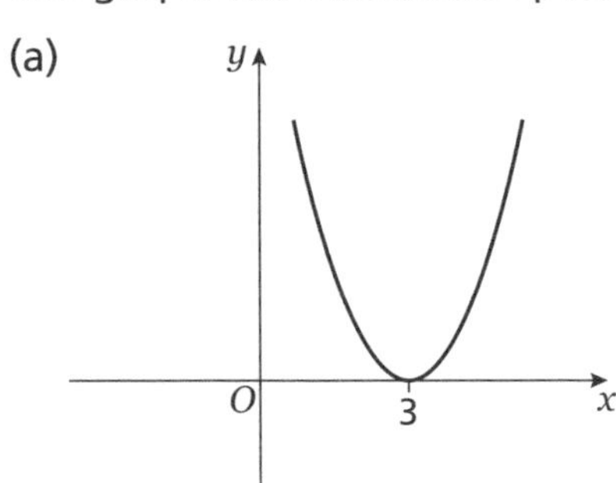

(b)

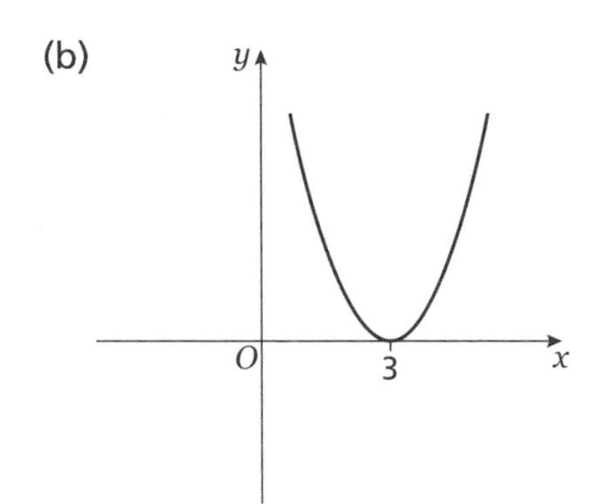

a On the axes (a) above, sketch the graph of $y = f(x + 2)$
Write down the coordinates of the new minimum point.

....

(2)

b On the axes (b) above, sketch the graph of $y = f(-x) + 2$
Write down the coordinates of the new minimum point.

....

(2)

(Total for Question 5 is 4 marks)

---

6 **a** Solve $x^2 - 6x + 7 = 0$
Give your answer in the form $a \pm \sqrt{b}$ where a and b are integers.

....

(3)

b The smaller root of $x^2 - 6x + 7 = 0$ is α
Find the value of α^3
Give your answer in the form $p - q\sqrt{r}$ where p, q and r are integers.

....

(3)

(Total for Question 6 is 6 marks)

---

7 **a** Simplify $\dfrac{2x^2 \times 3xy}{12x^4y}$

....

(2)

b Simplify $\left(\dfrac{3a^2b}{\sqrt{c}}\right)^{-2}$

....

(2)

c Solve $4^{2x} = 2^{x+1}$

....

(2)

(Total for Question 7 is 6 marks)

---

8 The roots of the quadratic equation $x^2 + bx + c = 0$ are 2 and -3

a Find the value of b and the value of c

$b =$ ....

$c =$ ....

(2)

The roots of $y^2 - 12y + 2q = 0$ are $y = \alpha$ and $y = 2\alpha$

b Find the value of α and the value of q

$\alpha =$ ....

$q =$ ....

(2)

(Total for Question 8 is 4 marks)

---

9 **a** Solve $4x - 3 > 7 - 2x$

....

(2)

y is an integer and $4y^2 < 64$

b Write down the possible values of y

....

(2)

(Total for Question 9 is 4 marks)

---

10 The first two terms of an arithmetic progression are 3 and 7

a Find the value of the 100th term.

....

(2)

Let S_n be the sum of the first n terms of the arithmetic progression with first term -20 and common difference 2

b Find the smallest value of n for which S_n is positive.

....

(3)

(Total for Question 10 is 5 marks)

---

11 A is the point with coordinates (2, 1).
B is the point with coordinates (6, 4).
C is the point with coordinates (3, 8).

a Show that angle ABC is a right angle.

(2)

The line through the points A and B cuts the y-axis at P.
The line through the points B and C cuts the y-axis at Q.

b Calculate the length of PQ.

....

(4)

(Total for Question 11 is 6 marks)

---

12 a Complete the table of values for $y = \text{f}(x) = x(x - 3)(x + 2)$

x	−3	−2	−1	0	1	2	3	4
y		0		0		−8		24

(1)

b On graph paper, draw the graph of $y = \text{f}(x) = x(x - 3)(x + 2)$ for values of x from −3 to 4

(3)

c Use the graph to estimate the values of x for which $\text{f}(x) = -4$

....

(2)

(Total for Question 12 is 6 marks)

---

13 a Simplify fully $\frac{x}{x^2 - 1} + \frac{3}{x + 1}$

....

(2)

b Solve $\frac{x}{x^2 - 1} + \frac{3}{x + 1} = 0$

....

(2)

(Total for Question 13 is 4 marks)

---

14 a Write down an expression, in terms of p, for the discriminant of the equation

$4x^2 + 4px + (4p + 5) = 0$

....

(1)

b Find the set of values of p for which the equation has two equal roots.

....

(3)

(Total for Question 14 is 4 marks)

15

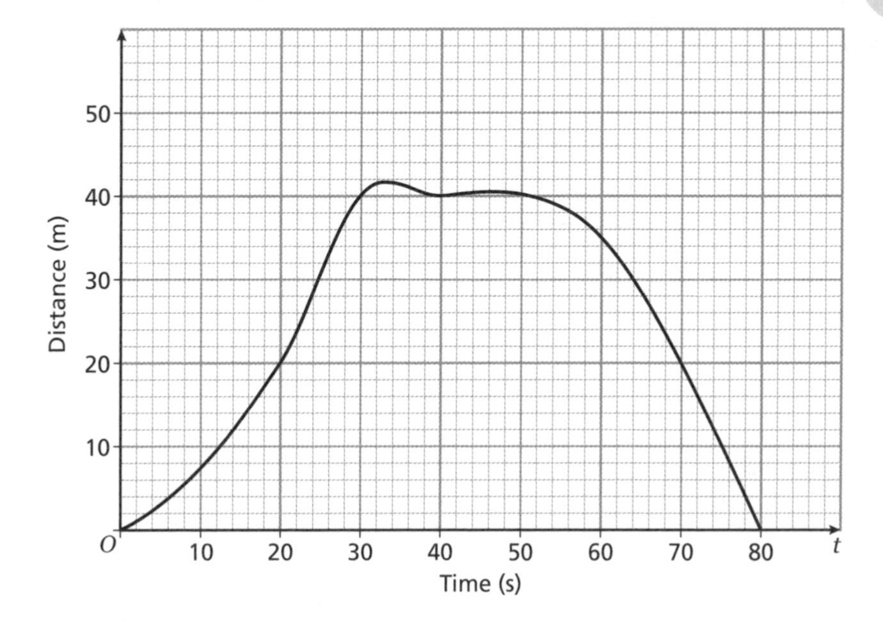

The graph shows the distance from O of a particle A moving in a straight line at time t seconds. Particle B starts from O at $t = 10$ and moves for 40 seconds in the same direction as A with a speed of 0.5 m/s. It then stops.

a Draw the distance–time graph of particle B.

(2)

b Write down the value of t at which particle A meets particle B.

....

(1)

c Find an estimate for the speed of particle A at $t = 70$

.... metres per second

(3)

(Total for Question 15 is 6 marks)

16 C is the circle with equation $(x - 3)^2 + (y + 2)^2 = 16$

a Write down the coordinates of the centre of C.

....

(1)

AB is the diameter of the circle that is parallel to the y-axis, with A above the x-axis.

b Find the coordinates of A and of B.

$A =$ ....

$B =$ ....

(3)

The point P (5, p) lies on C.

c Find the possible values of p.

....

(3)

(Total for Question 16 is 7 marks)

17 A firm makes oval plates of different sizes.
The mass, m grams, of each oval plate is proportional to the square of the transverse diameter, d cm, of the plate.
One size of plate has $d = 24$ and $m = 960$
Another size plate has $d = 18$

a Calculate the mass of this plate.

.... g
(4)

The circumference, C cm, of any oval plate the firm makes is directly proportional to the transverse diameter, d cm, of the plate.
$C = 36$ when $d = 12$

b Show that $C = 9\sqrt{\frac{m}{15}}$

(4)

(Total for Question 17 is 8 marks)

18 The straight line with equation $3y = 8x + 4$ meets the curve $y = 4 + 6x - x^2$ at the point A as shown in the sketch.

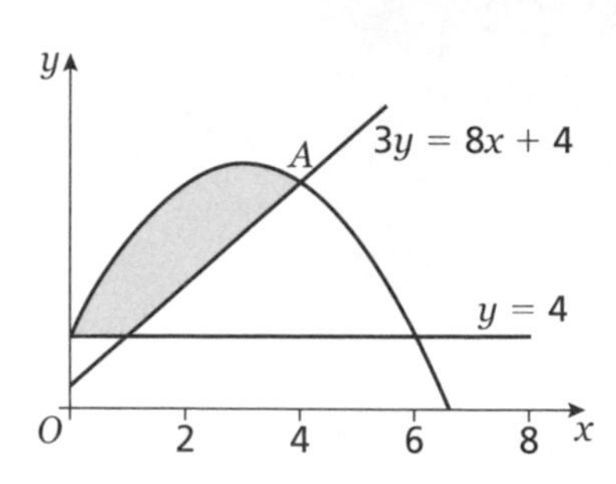

a Find the coordinates of the point A.

....

(4)

b Use the trapezium rule with 4 strips to find an estimate for the area of the region shown shaded in the sketch.

....

(5)

(Total for Question 18 is 9 marks)

TOTAL FOR PAPER IS 90 MARKS

Answers

1 Algebraic manipulation

1.1 Expanding two brackets

1 **a** $6x^2 - 15x$

b

$\times$	x	$+2$
x	x^2	$+2x$
$+3$	$+3x$	$+6$

$x^2 + 2x + 3x + 6 = x^2 + 5x + 6$

c $2x^2 - 10x + 3x - 15 = 2x^2 - 7x - 15$

d $6x^2 - 8xy - 15xy + 20y^2 = 6x^2 - 23xy + 20y^2$

2 **a** $2x^2 + 8x$ **b** $18x^2 - 30x$
c $10x^2 - 10xy$ **d** $x^2 + 9x + 20$
e $x^2 + 10x + 21$ **f** $x^2 + 5x - 14$
g $x^2 - 25$ **h** $2x^2 + x - 3$
i $6x^2 - x - 2$ **j** $10x^2 - 31x + 15$
k $12x^2 + 13x - 14$ **l** $18x^2 + 39xy + 20y^2$
m $35x^2 + 14xy - 15x - 6y$ **n** $6x^2 - 16x - 9xy + 24y$

1.2 Factorising expressions

1 **a** $3x^2y(5y^2 + 3x^2)$ **b** $(2x - 5y)(2x + 5y)$

c $b = 3, ac = -10$
$x^2 + 3x - 10 = x^2 + 5x - 2x - 10$
$= x(x + 5) - 2(x + 5)$
$= (x + 5)(x - 2)$

d $b = -11, ac = -60$
Two numbers are -15 and 4
$6x^2 - 11x - 10 = 6x^2 - 15x + 4x - 10$
$= 3x(2x - 5) + 2(2x - 5)$
$= (3x + 2)(2x - 5)$

2 **a** $2x^3y^3(3x - 5y)$ **b** $7a^3b^2(3b^3 + 5a^2)$ **c** $5x^2y^2(5 - 2x + 3y)$

3 **a** $(x + 3)(x + 4)$ **b** $(x + 7)(x - 2)$ **c** $(x - 5)(x - 6)$
d $(x - 8)(x + 3)$ **e** $(x - 9)(x + 2)$ **f** $(x + 5)(x - 4)$
g $(x - 8)(x + 5)$ **h** $(x + 7)(x - 4)$

4 **a** $(6x - 7y)(6x + 7y)$ **b** $(2x - 9y)(2x + 9y)$
c $2(3a - 10bc)(3a + 10bc)$

5 **a** $(x - 1)(2x + 3)$ **b** $(3x + 1)(2x + 5)$ **c** $2(3x - 2)(2x - 5)$
d $(2x + 1)(x + 3)$ **e** $(3x - 1)(3x - 4)$ **f** $(5x + 3)(2x + 3)$

1.3 Using index laws

1 **a** 1 **b** $\sqrt{9} = 3$ **c** $(\sqrt[3]{27})^2 = 3^2 = 9$
d $\frac{1}{4}^2 = \frac{1}{16}$ **e** $3x^3$ **f** $\frac{x^8}{x^4} = x^4$
g $\frac{x^5}{x^{\frac{5}{2}}} = x^{\frac{5}{2}}$ **h** $\frac{12x^2}{8x^6} = \frac{3}{2x^4}$

2 **a** 1 **b** 1 **c** 1

3 **a** 7 **b** 4 **c** 5 **d** 2

4 **a** 125 **b** 32 **c** 343 **d** 8

5 **a** $\frac{1}{25}$ **b** $\frac{1}{64}$ **c** $\frac{1}{32}$ **d** $\frac{1}{36}$

6 **a** $\frac{3x^3}{2}$ **b** $5x^2$ **c** $3x$ **d** $\frac{y}{2x^2}$ **e** $y^{\frac{1}{2}}$ **f** c^{-3}
g $2x^6$ **h** x

7 **a** $\frac{1}{2}$ **b** $\frac{1}{9}$ **c** $\frac{8}{3}$ **d** $\frac{1}{4}$ **e** $\frac{4}{3}$ **f** $\frac{16}{9}$

1.4 Algebraic fractions

1 **a** $\frac{2x(x - 2)}{6x(2 + x)} = \frac{x - 2}{3(x + 2)}$

b $\frac{(x + 3)(x - 7)}{(2x + 3)(x + 3)} = \frac{x - 7}{2x + 3}$

c $\frac{2x}{6} + \frac{6x + 3}{6} = \frac{8x + 3}{6}$

d $\frac{2x + 2}{(x - 3)(x + 1)} - \frac{5x - 15}{(x - 3)(x + 1)} = \frac{-3x + 17}{(x - 3)(x + 1)}$

2 **a** $\frac{2(x + 2)}{x - 1}$ **b** $\frac{x}{x - 1}$ **c** $\frac{x + 2}{x}$
d $\frac{x}{x + 5}$ **e** $\frac{x + 3}{x}$ **f** $\frac{x}{x - 5}$

3 **a** $\frac{13x}{15}$ **b** $\frac{11x + 5}{10}$ **c** $\frac{x}{28}$
d $\frac{x}{12}$ **e** $\frac{11x + 4}{12}$ **f** $\frac{7x + 13}{20}$

4 **a** $\frac{5x + 11}{(x + 3)(x + 1)}$ **b** $\frac{3(x + 1)}{x(x + 3)}$ **c** $\frac{x - 8}{x(x + 4)}$
d $\frac{2(x - 3)}{(x + 1)(x - 1)}$ **e** $\frac{5(x + 2)}{(2x - 3)(x + 1)}$ **f** $\frac{5x - 4}{(x + 1)(x - 2)}$

5 **a** $\frac{3x + 4}{x + 7}$ **b** $\frac{2x + 3}{3x - 2}$ **c** $\frac{2 - 5x}{2x - 3}$ **d** $\frac{3x + 1}{x + 4}$

1.5 Completing the square

1 **a** $(x + 3)^2 - 2 - 9 = (x + 3)^2 - 11$

b $2(x^2 - \frac{5}{2}x + \frac{1}{2}) = 2[(x - \frac{5}{4})^2 + \frac{1}{2} - \frac{25}{16}] = 2[(x - \frac{5}{4})^2 - \frac{17}{16}]$
$= 2(x - \frac{5}{4})^2 - \frac{17}{8}$

2 **a** $(x + 2)^2 - 1$ **b** $(x - 5)^2 - 28$ **c** $(x - 4)^2 - 16$
d $(x + 3)^2 - 9$ **e** $(x - 1)^2 + 6$ **f** $(x + \frac{3}{2})^2 - \frac{17}{4}$

3 **a** $2(x - 2)^2 - 24$ **b** $4(x - 1)^2 - 20$
c $3(x + 2)^2 - 21$ **d** $2(x + \frac{3}{2})^2 - \frac{25}{2}$

4 **a** $2(x + \frac{3}{4})^2 + \frac{39}{8}$ **b** $3(x - \frac{1}{3})^2 - \frac{1}{3}$
c $5(x + \frac{3}{10})^2 - \frac{9}{20}$ **d** $3(x + \frac{5}{6})^2 + \frac{11}{12}$

Don't forget!

* four
* $ax^2 + bx + c$
* b; ac
* the difference of two squares; $(x - y)(x + y)$
* a^{m+n}
* a^{m-n}
* a^{mn}
* 1
* $\sqrt[n]{a}$
* $\sqrt[n]{(a^m)}$ or $(\sqrt[n]{a})^m$
* $\frac{1}{a^m}$
* numerator; denominator
* 1
* common denominator; equivalent
* $p(x + q)^2 + r$

Exam-style questions

1 **a** $3x^2 - 7x - 6$ **b** $6x^2y^2(2x + 5y^3)$ **c** x^2

2 **a** x^{-2} **b** $(x - 5)(x + 7)$ **c** $(2x - 5y)(2x + 5y)$

3 $(x + 1\frac{1}{2})^2 - 7\frac{1}{4}$

4 $\frac{x + 2}{2x + 3}$

2 Formulae

2.1 Substitution

1 **a** $2 \times 8 + (-6) = 16 - 6 = 10$ **b** $8 + (-6) \times \frac{1}{3} = 8 - 2 = 6$
c $\frac{3 \times 8}{-6} = -4$ **d** $8^{\frac{1}{3}} - (-6) = 2 + 6 = 8$

2 $C = \frac{5}{9}$ of $(50 - 32)$
$C = \frac{5}{9}$ of 18
$C = 5 \times 18 \div 9$
$C = 10$

3 a 7 b 3 c $20\frac{1}{2}$ d 25
e -324 f -18

4 a -1.6 b 2.7 c 2.8 d -2.4

5 a $-2\frac{2}{3}$ b $-\frac{1}{6}$ c $-1\frac{1}{6}$ d -7

6 610

2.2 Changing the subject of a formula

1 $v - u = at$
$t = \dfrac{v-u}{a}$

2 $r = t(2 - \pi)$
$t = \dfrac{r}{2-\pi}$

3 $2(t + r) = 5 \times 3t$
$2t + 2r = 15t$
$2r = 13t$
$t = \dfrac{2r}{13}$

4 $r(t - 1) = 3t + 5$
$rt - r = 3t + 5$
$rt - 3t = 5 + r$
$t(r - 3) = 5 + r$
$t = \dfrac{5+r}{r-3}$

5 $d = \dfrac{C}{\pi}$ 6 $w = \dfrac{P-2l}{2}$ 7 $T = \dfrac{S}{D}$

8 $t = \dfrac{q-r}{p}$ 9 $t = \dfrac{2u}{2a-1}$ 10 $x = \dfrac{V}{a+4}$

11 $y = 2 + 3x$ 12 $a = \dfrac{3x+1}{x+2}$ 13 $d = \dfrac{b-c}{a}$

14 $g = \dfrac{2h+9}{7-h}$ 15 $e = \dfrac{1}{x+7}$

Don't forget!

* replacing each letter with its value
* everything else

Exam-style questions

1 $x = \dfrac{4y-3}{2+y}$

3 Surds

3.1 Surds

1 $\sqrt{25 \times 2} = \sqrt{25} \times \sqrt{2} = 5 \times \sqrt{2} = 5\sqrt{2}$

2 $\sqrt{49 \times 3} - 2\sqrt{4 \times 3} = \sqrt{49} \times \sqrt{3} - 2\sqrt{4} \times \sqrt{3}$
$= 7 \times \sqrt{3} - 2 \times 2 \times \sqrt{3} = 3\sqrt{3}$

3 $\sqrt{49} - \sqrt{7}\sqrt{2} + \sqrt{2}\sqrt{7} - \sqrt{4} = 7 - 2 = 5$

4 a $3\sqrt{5}$ b $5\sqrt{5}$ c $4\sqrt{3}$ d $5\sqrt{7}$
e $10\sqrt{3}$ f $2\sqrt{7}$ g $6\sqrt{2}$ h $9\sqrt{2}$

5 a -1 b $9 - \sqrt{3}$ c $10\sqrt{5} - 7$ d $26 - 4\sqrt{2}$

6 a $15\sqrt{2}$ b $\sqrt{5}$ c $3\sqrt{2}$ d $\sqrt{3}$
e $6\sqrt{7}$ f $5\sqrt{3}$

3.2 Rationalising the denominator

1 a $\dfrac{1}{\sqrt{3}} \times \dfrac{\sqrt{3}}{\sqrt{3}} = \dfrac{\sqrt{3}}{3}$

b $\dfrac{\sqrt{2}}{\sqrt{12}} \times \dfrac{\sqrt{12}}{\sqrt{12}} = \dfrac{\sqrt{2} \times 2\sqrt{3}}{12} = \dfrac{\sqrt{6}}{6}$

c $\dfrac{3}{2+\sqrt{5}} \times \dfrac{2-\sqrt{5}}{2-\sqrt{5}} = \dfrac{3(2-\sqrt{5})}{4 + 2\sqrt{5} - 2\sqrt{5} - 5} = \dfrac{3(2-\sqrt{5})}{-1}$
$= -3(2 - \sqrt{5}) = -6 + 3\sqrt{5}$

2 a $\dfrac{\sqrt{5}}{5}$ b $\dfrac{\sqrt{11}}{11}$ c $\dfrac{2\sqrt{7}}{7}$ d $\dfrac{\sqrt{2}}{2}$
e $\sqrt{2}$ f $\sqrt{5}$ g $\dfrac{\sqrt{3}}{3}$ h $\frac{1}{3}$

3 a $\dfrac{3+\sqrt{5}}{4}$ b $\dfrac{2(4-\sqrt{3})}{13}$ c $\dfrac{6(5+\sqrt{2})}{23}$

Don't forget!

* the square root of a number that is not a square number
* $\sqrt{2}, \sqrt{3}, \sqrt{5}$, etc.
* $\sqrt{a} \times \sqrt{b}$
* $\dfrac{\sqrt{a}}{\sqrt{b}}$
* denominator
* $\sqrt{b}$
* $b - \sqrt{c}$

Exam-style questions

1 $2\sqrt{5}$ 2 $9 - 4\sqrt{2}$ 3 $10 + 5\sqrt{3}$ 4 $\dfrac{3\sqrt{5}}{5}$ 5 $7\sqrt{2}$

4 Quadratic equations

4.1 Solving by factorisation

1 a $5x^2 - 15x = 0$
$5x(x - 3) = 0$
So $5x = 0$ or $x - 3 = 0$
$x = 0$ or $x = 3$

b $(x + 4)(x + 3) = 0$
So $x + 4 = 0$ or $x + 3 = 0$
$x = -4$ or $x = -3$

c $(3x + 4)(3x - 4) = 0$
So $3x + 4 = 0$ or $3x - 4 = 0$
$x = -1\frac{1}{3}$ or $x = 1\frac{1}{3}$

d $(2x + 3)(x - 4) = 0$
So $2x + 3 = 0$ or $x - 4 = 0$
$x = -1\frac{1}{2}$ or $x = 4$

2 a $x = 0$ or $x = -\frac{2}{3}$ b $x = 0$ or $x = \frac{3}{4}$
c $x = -5$ or $x = -2$ d $x = 2$ or $x = 3$
e $x = -1$ or $x = 4$ f $x = -5$ or $x = 2$
g $x = 4$ or $x = 6$ h $x = -6$ or $x = 6$
i $x = -7$ or $x = 4$ j $x = 3$
k $x = -\frac{1}{2}$ or $x = 4$ l $x = -\frac{2}{3}$ or $x = 5$

3 a $x = -2$ or $x = 5$ b $x = -1$ or $x = 3$
c $x = -8$ or $x = 3$ d $x = -6$ or $x = 7$
e $x = -5$ or $x = 5$ f $x = -4$ or $x = 7$
g $x = -3$ or $x = 2\frac{1}{2}$ h $x = -\frac{1}{3}$ or $x = 2$

4.2 Solving by completing the square

1 $(x + 3)^2 + 4 - 9 = 0$
$(x + 3)^2 - 5 = 0$
$(x + 3)^2 = 5$
$x + 3 = \pm\sqrt{5}$
$x = -3 \pm\sqrt{5}$
$x = -3 + \sqrt{5}$ or $x = -3 - \sqrt{5}$

2 $2[x^2 - \frac{7}{2}x + 2] = 0$
$2[(x - \frac{7}{4})^2 + 2 - \frac{49}{16}] = 0$
$(x - \frac{7}{4})^2 - \frac{17}{16} = 0$
$(x - \frac{7}{4})^2 = \frac{17}{16}$
$x - \frac{7}{4} = \pm\sqrt{\dfrac{17}{16}}$
$x - \frac{7}{4} = \pm\frac{1}{4}\sqrt{17}$
$x = \dfrac{7+\sqrt{17}}{4}$ or $x = \dfrac{7-\sqrt{17}}{4}$

3 a $x = 2 + \sqrt{7}$ or $x = 2 - \sqrt{7}$
b $x = 5 + \sqrt{21}$ or $x = 5 - \sqrt{21}$
c $x = -4 + \sqrt{21}$ or $x = -4 - \sqrt{21}$
d $x = 1 + \sqrt{7}$ or $x = 1 - \sqrt{7}$
e $x = -2 + \sqrt{6.5}$ or $x = -2 - \sqrt{6.5}$
f $x = \dfrac{-3+\sqrt{89}}{10}$ or $x = \dfrac{-3-\sqrt{89}}{10}$

4 a $x = 1 + \sqrt{14}$ or $x = 1 - \sqrt{14}$
b $x = \dfrac{-3+\sqrt{23}}{2}$ or $x = \dfrac{-3-\sqrt{23}}{2}$
c $x = \dfrac{5+\sqrt{13}}{2}$ or $x = \dfrac{5-\sqrt{13}}{2}$

4.3 Solving by using the formula

1 $x = \dfrac{-(6) \pm \sqrt{(6)^2 - 4 \times 1 \times 4}}{2 \times 1}$
$x = \dfrac{-6 \pm \sqrt{36 - 16}}{2}$
$x = \dfrac{-6 \pm \sqrt{20}}{2}$
$x = \dfrac{-6 \pm \sqrt{4 \times 5}}{2}$
$x = \dfrac{-6 + 2\sqrt{5}}{2}$ or $x = \dfrac{-6 - 2\sqrt{5}}{2}$
$x = -3 + \sqrt{5}$ or $x = -3 - \sqrt{5}$

2 $a = 3, b = -7, c = -2$

$x = \frac{7 \pm \sqrt{49 - 4 \times 3 \times -2}}{2 \times 3}$

$x = \frac{7 \pm \sqrt{49 + 24}}{6}$

$x = \frac{7 \pm \sqrt{73}}{6}$

$x = \frac{7 + \sqrt{73}}{6}$ or $x = \frac{7 - \sqrt{73}}{6}$

3 **a** $x = -1 + \frac{\sqrt{3}}{3}$ or $x = -1 - \frac{\sqrt{3}}{3}$

b $x = 1 + \frac{3\sqrt{2}}{2}$ or $x = 1 - \frac{3\sqrt{2}}{2}$

4 **a** $x = \frac{7 + \sqrt{17}}{8}$ or $x = \frac{7 - \sqrt{17}}{8}$

b $x = -1 + \sqrt{10}$ or $x = -1 - \sqrt{10}$

Don't forget!

* two; b; ac
* $\frac{-b \pm \sqrt{b^2 - 4ac}}{2a}$
* negative

Exam-style questions

1 $x = \frac{7 + \sqrt{41}}{2}$ or $x = \frac{7 - \sqrt{41}}{2}$

2 $x = -1\frac{2}{3}$ or $x = 2$

3 $x = \frac{-3 + \sqrt{89}}{20}$ or $x = \frac{-3 - \sqrt{89}}{20}$

5 Roots of quadratic equations

5.1 The role of the discriminant

1 $a = 3, b = 7, c = 5$

$b^2 - 4ac = 7^2 - 4 \times 3 \times 5 = 49 - 60 = -11$; no real roots

2 $b^2 - 4ac = 0$

$a = 1, b = 4, c = p$

$b^2 - 4ac = 4^2 - 4 \times 1 \times p$

$16 - 4p = 0$

$4p = 16$

$p = 4$

3 $b^2 - 4ac < 0$

$a = h, b = 3, c = -7$

$b^2 - 4ac = 3^2 - 4 \times h \times -7 = 9 + 28h$

$9 + 28h < 0$

$28h < -9$

$h < -\frac{9}{28}$

4 no real roots

5 two real and distinct roots

6 two real and equal roots

7 no real roots

8 $q = \pm 8$

9 $q = \pm 6\sqrt{2}$

10 $r = \pm 5$

11 $t > -\frac{4}{3}$

5.2 The sum and product of the roots of a quadratic equation

1 $a = 2, b = 6, c = -5$

Sum $= -\frac{b}{a} = -\frac{6}{2} = -3$

Product $= \frac{c}{a} = \frac{-5}{2} = -2.5$

2 $-\frac{b}{a} = -7, \frac{c}{a} = 10$

$x^2 - (-7)x + (10) = 0$

$x^2 + 7x + 10 = 0$

3 sum = 11, product = 30

4 sum = -1.6, product = -4.2

5 sum = 0, product = $\frac{-16}{9}$

6 sum = $\frac{-1}{6}$, product = $\frac{-5}{2}$

7 $x^2 + 2x - 8 = 0$

8 $3x^2 + x - 2 = 0$

9 $2x^2 + 17x - 9 = 0$

10 $2x^2 + 3x - 29 = 0$

Don't forget!

* $x = \frac{-b \pm \sqrt{b^2 - 4ac}}{2a}$
* discriminant
* two real and distinct roots
* two real and equal roots
* no real roots
* $-\frac{b}{a}$
* $\frac{c}{a}$
* $x^2 - (-\frac{b}{a})x + \frac{c}{a} = 0$

Exam-style questions

1 $\pm 8\sqrt{3}$

2 $2x^2 + 5x + 9 = 0$

6 Simultaneous equations

6.1 Solving simultaneous linear equations using elimination

1 $2x = 4$

$x = 2, y = -1$

2 $6x = 18$

$x = 3, y = 5$

3 $x = 1, y = 4$

4 $x = 3, y = -2$

5 $x = 2, y = -5$

6 $x = 3, y = -\frac{1}{2}$

7 $x = 6, y = -1$

8 $x = -2, y = 5$

6.2 Solving simultaneous linear equations using substitution

1 $5x + 3(2x + 1) = 14$

$5x + 6x + 3 = 14$

$11x = 11$

$x = 1, y = 3$

2 $4x + 3(2x - 16) = -3$

$4x + 6x - 48 = -3$

$10x = 45$

$x = 4.5, y = -7$

3 $x = 9, y = 5$

4 $x = -2, y = -7$

5 $x = \frac{1}{2}, y = 3\frac{1}{2}$

6 $x = \frac{1}{2}, y = 3$

7 $x = -4, y = 5$

8 $x = -2, y = -5$

9 $x = \frac{1}{4}, y = 1\frac{3}{4}$

10 $x = -2, y = \frac{5}{2}$

6.3 Solving simultaneous equations where one is quadratic

1 $x^2 + (x + 1)^2 = 13$

$x^2 + x^2 + 2x + 1 - 13 = 0$

$2x^2 + 2x - 12 = 0$

$2(x^2 + x - 6) = 0$

$(x + 3)(x - 2) = 0$

$x = -3$ or $x = 2$

when $x = -3, y = -2$

when $x = 2, y = 3$

2 $x = \frac{5 - 3y}{2}$

$2y^2 + \frac{y(5 - 3y)^2}{2} = 12$

$2y^2 + \frac{5y - 3y^2}{2} - 12 = 0$

$4y^2 + 5y - 3y^2 - 24 = 0$

$y^2 + 5y - 24 = 0$

$(y + 8)(y - 3) = 0$

$y = -8$ or $y = 3$

when $y = -8, x = 14\frac{1}{2}$

when $y = 3, x = -2$

3 $x = 0, y = 5$

$x = -5, y = 0$

4 $x = -\frac{8}{3}, y = -\frac{19}{3}$

$x = 3, y = 5$

5 $x = -2, y = -4$

$x = 2, y = 4$

6 $x = \frac{5}{2}, y = 6$

$x = 3, y = 5$

7 $x = \frac{1 + \sqrt{5}}{2}, y = \frac{-1 + \sqrt{5}}{2}$

$x = \frac{1 - \sqrt{5}}{2}, y = \frac{-1 - \sqrt{5}}{2}$

8 $x = \frac{-1 + \sqrt{7}}{2}, y = \frac{3 + \sqrt{7}}{2}$

$x = \frac{-1 - \sqrt{7}}{2}, y = \frac{3 - \sqrt{7}}{2}$

Don't forget!

* elimination; substitution
* two

Exam-style questions

1 $x = 4, y = -2$

$x = -3\frac{1}{2}, y = 2\frac{1}{2}$

2 $x = 2\frac{1}{2}, y = \frac{1}{2}$

7 Arithmetic series

7.1 General (nth) term of arithmetic series

1 First term $= 4 \times 1 + 1 = 5$

Second term $= 4 \times 2 + 1 = 9$

Third term $= 4 \times 3 + 1 = 13$

Fourth term $= 4 \times 4 + 1 = 17$

Fifth term $= 4 \times 5 + 1 = 21$

First 5 terms are 5, 9, 13, 17, 21

2 $5n - 2 = 73$
$5n = 75$
$n = 15$

3 $a = 3, d = 5$
nth term $= 3 + (n - 1) \times 5$
$= 3 + 5n - 5$
$= 5n - 2$

4 $8 + 13 + 18$ **5** $3n + 2; 62$ **6** $17 - 2n; -3$ **7** $82; 402$
8 $8; -97$ **9** 25 **10** 53 **11** 5
12 first term = 1, common difference = 3

7.2 The sum of an arithmetic series

1 $a = 1, d = 4, n = 30$
$S_n = \frac{30}{2}[2 \times 1 + (30 - 1) \times 4]$
$S_n = 15 \times (2 + 29 \times 4)$
$S_n = 1770$

2 $S_n = 432, a = 7, L = 41$
$432 = \frac{n}{2}(7 + 41)$
$432 = 24n$
$n = 18$

3 $S_n = 352, a = 7, d = 2$
$352 = \frac{n}{2}[2 \times 7 + (n - 1) \times 2]$
$704 = n(14 + 2n - 2)$
$704 = 2n^2 + 12n$
$2n2 + 12n - 704 = 0$
$n^2 + 6n - 352 = 0$
$(n + 22)(n - 16) = 0$
$n = 16$

4 $S_n = 1, n = 2$
$1 = \frac{2}{2}[2a + (2 - 1)d]$
$2a + d = 1$
$93 = a + (20 - 1)d$
$a + 19d = 93$
$a + 19(1 - 2a) = 93$
$a + 19 - 38a = 93$
$19 - 37a = 93$
$-37a = 74$
$a = -2, d = 5$
first term = −2; common difference = 5

5 610 **6** 1395 **7** −5350 **8** 290
9 341 **10** 1370 **11** 488 **12** 10
13 first term = 2; common difference = 3

Don't forget!

* sequence
* nth term
* the same amount
* $a + (n - 1)d$
* $\frac{n}{2}[2a + (n - 1)d]$
* $\frac{n}{2}(a + L)$

Exam-style questions

1 a first term = 60; common difference = −7
b −1245

8 Coordinate geometry

8.1 The equation of a line

1 $y = -\frac{1}{2}x + 3$
$2y = -x + 6$
$x + 2y - 6 = 0$

2 $3y = 2x - 4$
$y = \frac{2}{3}x - \frac{4}{3}$
gradient $= m = \frac{2}{3}$
y-intercept $= c = -\frac{4}{3}$ or $-1\frac{1}{3}$

3 $m = 3$
$y = 3x + c$
$13 = 3 \times 5 + c$
$13 = 15 + c$
$c = -2$
$y = 3x - 2$

4 $m = \frac{7 - 4}{8 - 2} = \frac{3}{6} = \frac{1}{2}$
$y = \frac{1}{2}x + c$
$4 \text{ (or 7)} = \frac{1}{2} \times 2 \text{ (or 8)} + c$
$4 \text{ (or 7)} = 1 \text{ (or 4)} + c$
$c = 3$
$y = \frac{1}{2}x + 3$

5 a $m = 3, c = 5$ **b** $m = -\frac{1}{2}, c = -7$
c $m = 2, c = -\frac{3}{2}$ **d** $m = -1, c = 5$
e $m = \frac{2}{3}, c = -\frac{7}{3}$ or $-2\frac{1}{3}$ **f** $m = -5, c = 4$

6 $y = 5x$
$y = -3x + 2$
$y = 4x - 7$

7 a $x + 2y + 14 = 0$ **b** $2x - y = 0$
c $2x - 3y + 12 = 0$ **d** $6x + 5y + 10 = 0$

8 $y = 4x - 3$

9 $y = -\frac{2}{3}x + 7$

10 a $y = 2x - 3$ **b** $y = -\frac{1}{2}x + 6$
c $y = 5x - 2$ **d** $y = -3x + 19$

8.2 Parallel and perpendicular lines

1 $m = 2$
$y = 2x + c$
$9 = 2 \times 4 + c$
$c = 1$
$y = 2x + 1$

2 $m = 2$
$-\frac{1}{m} = -\frac{1}{2}$
$y = -\frac{1}{2}x + c$
$5 = -\frac{1}{2} \times -2 + c = 1 + c$
$c = 4$
$y = -\frac{1}{2}x + 4$

3 $m = \frac{1}{2}$
$-\frac{1}{m} = -2$
$y = -2x + c$
$3 = -2 \times -5 + c$
$c = -7$
$y = -2x - 7$

4 $m = \frac{-1 - 5}{9 - 0} = -\frac{6}{9} = -\frac{2}{3}$
$-\frac{1}{m} = \frac{3}{2}$
$y = \frac{3}{2}x + c$
$\left(\frac{0 + 9}{2}, \frac{5 + -1}{2}\right) = \left(\frac{9}{2}, 2\right)$
$y = \frac{3}{2}x + c$
$2 = \frac{3}{2} \times \frac{9}{2} + c = \frac{27}{4} + c$
$c = -\frac{19}{4}$
$y = \frac{3}{2}x - \frac{19}{4}$

5 a $y = 3x - 7$ **b** $y = -2x + 5$
c $y = -\frac{1}{2}x$ **d** $y = \frac{3}{2}x + 8$

6 a $y = -\frac{1}{2}x + 2$ **b** $y = 3x + 7$
c $y = -4x + 35$ **d** $y = \frac{5}{2}x - 8$

7 a $y = -\frac{1}{2}x$ **b** $y = 2x$

8 a parallel **b** neither **c** perpendicular
d perpendicular **e** neither **f** parallel

Don't forget!

* $y = mx + c$
* $ax + by + c = 0$
* $m = \frac{y_2 - y_1}{x_2 - x_1}$
* gradient
* $-\frac{1}{m}$

Exam-style questions

1 a $x + 2y - 4 = 0$ **b** $x + 2y + 2 = 0$ **c** $y = 2x$

9 Graphs of functions

9.1 Recognising graphs

1 a, b

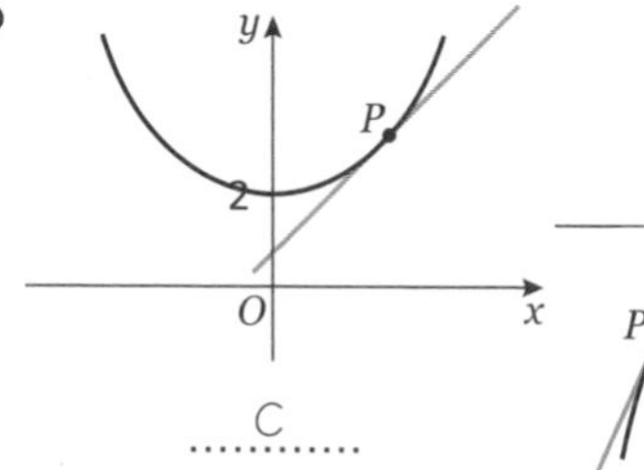

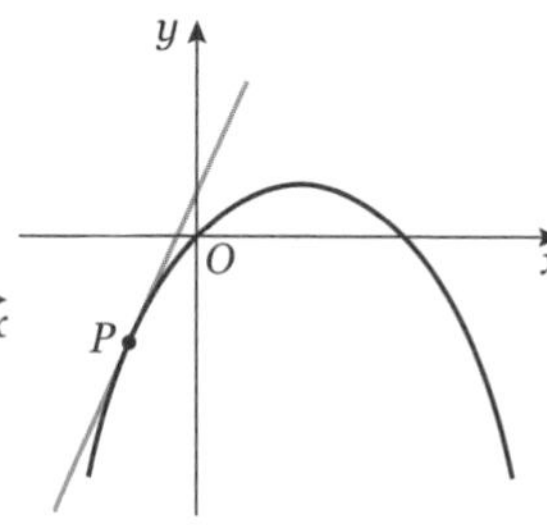

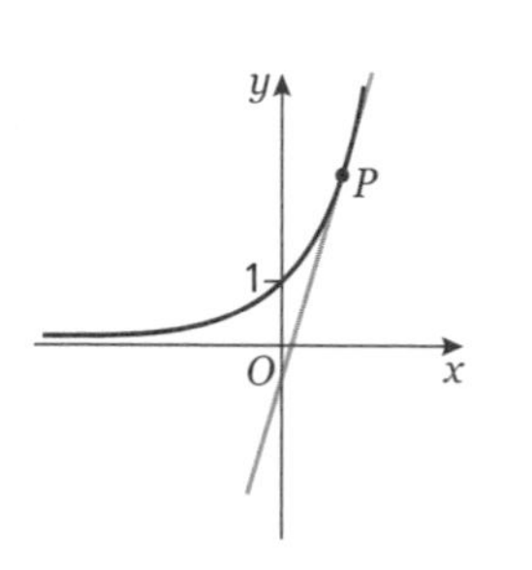

2 a, b

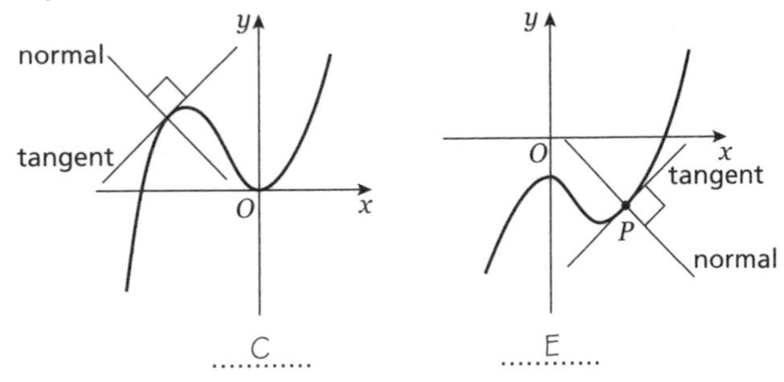

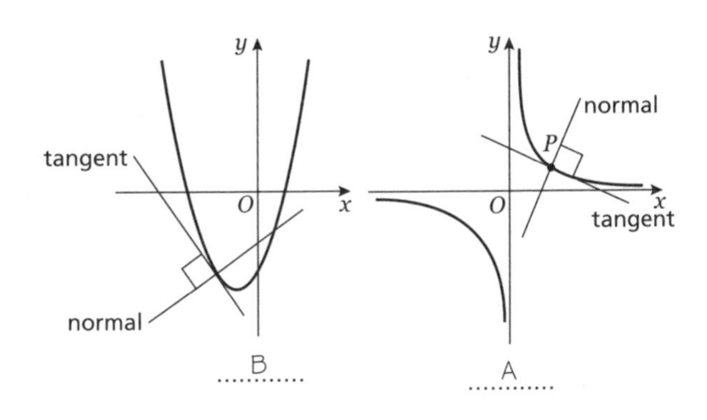

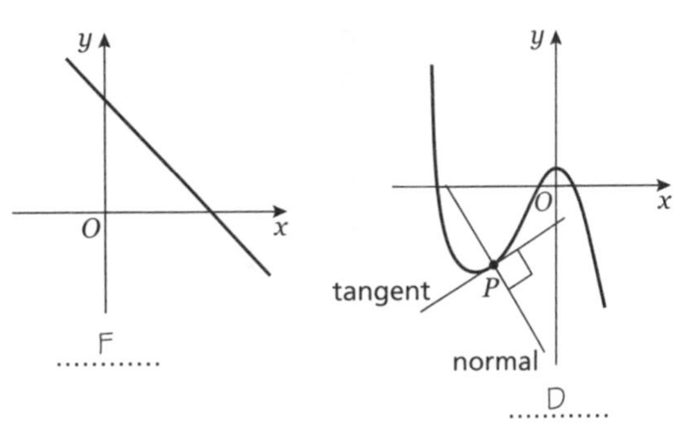

3 B, C, A

9.2 Drawing and using graphs

1 a

x	−3	−2	−1	0	1	2	3	4	5
y	10	3	−2	−5	−6	−5	−2	3	10

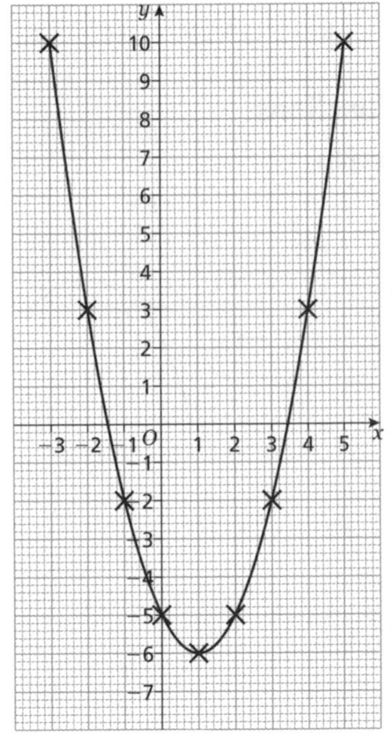

b $x \approx -1.4$ or -1.5 or $x \approx 3.4$ or 3.5

2 a

x	−2	−1	0	1	2	3	4
y	−13	−3	3	5	3	−3	−13

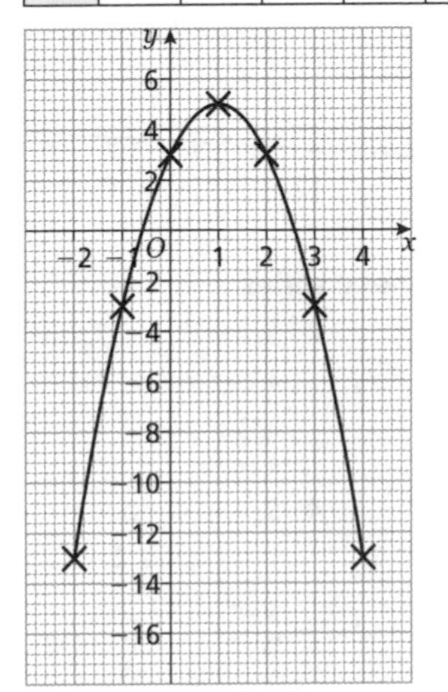

b $x \approx -0.6$ or $x \approx 2.6$

3 a

x	−7	−6	−5	−4	−3	−2	−1	0	1	2
y	11	3	−3	−7	−9	−9	−7	−3	3	11

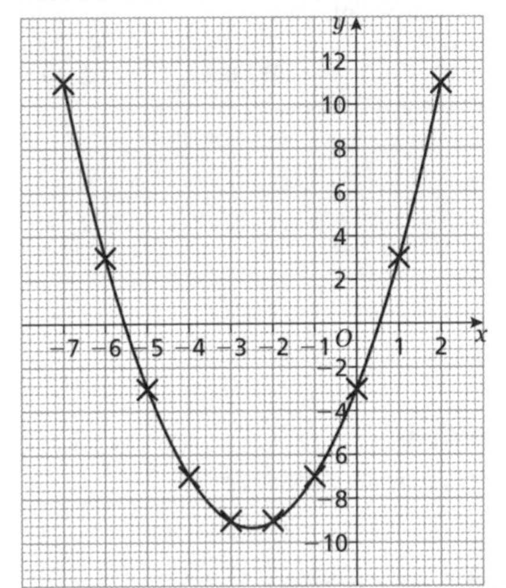

b $x \approx -5.5$ or -5.6 or $x \approx 0.5$ or 0.6

4 a

x	−3	−2	−1	0	1	2	3
y	−18	−1	4	3	2	7	24

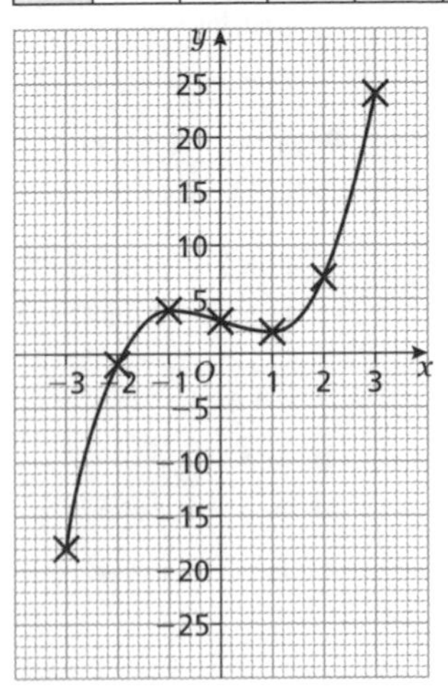

b $x \approx -1.4$, $x = 0$ or $x \approx 1.4$

5 a

x	−3	−2	−1	0	1	2	3
y	13	2	1	4	5	−2	−23

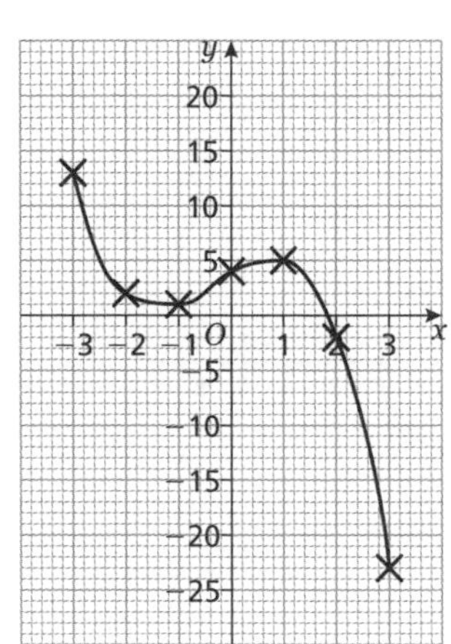

b $x \approx 1.8$

c $x = -2, x = -1$ or $x = 2$

6 a, b

x	−1	0	1	2	3	4	5
y	5	0	−3	−4	−3	0	5

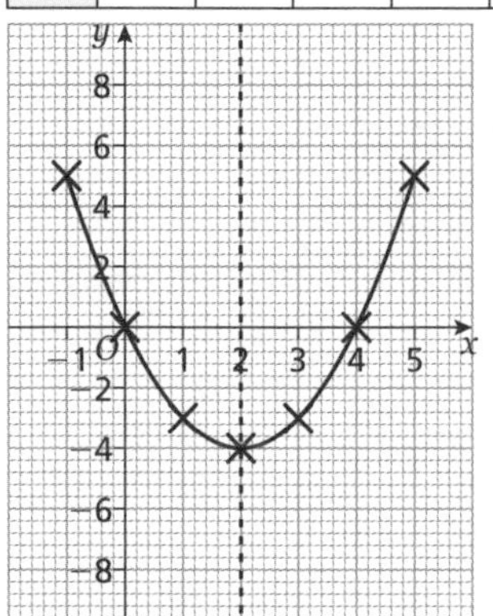

b $x = 2$

9.3 Sketching graphs

1

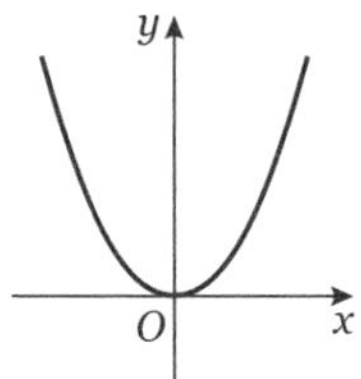

2 When $x = 0, y = 0^2 - 0 - 6 = -6$

$(0, -6)$

$(x - 3)(x + 2) = 0$

$x = 3$ or $x = -2$

$(-2, 0)$ and $(3, 0)$

$(x - \frac{1}{2})^2 - 6\frac{1}{4}$

$x = \frac{1}{2}$ and $y = -6\frac{1}{4}$

$(\frac{1}{2}, -6\frac{1}{4})$

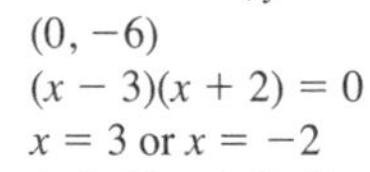

3

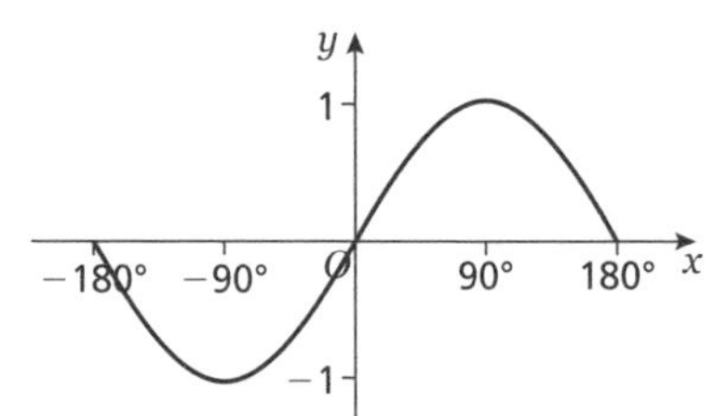

4 $x = 3, 1$ or -2

$(-2, 0), (1, 0)$ and $(3, 0)$

$y = -3 \times -1 \times 2 = 6$

$(0, 6)$

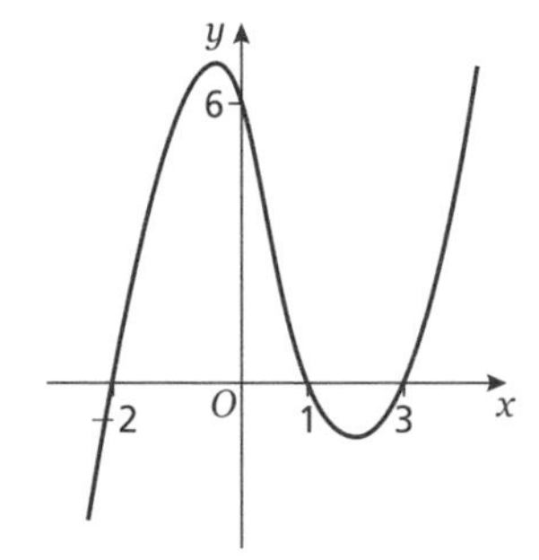

5

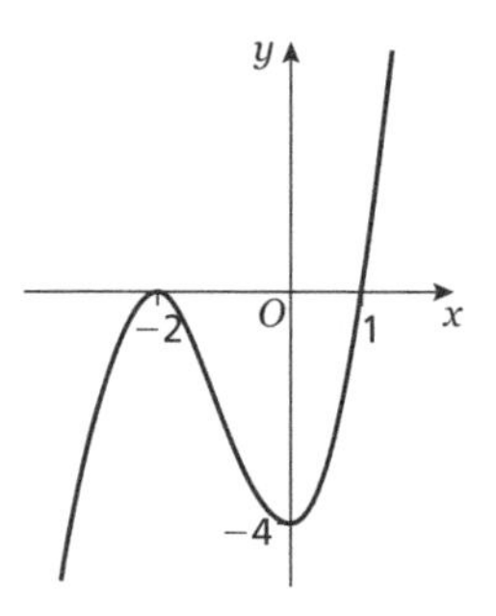

6

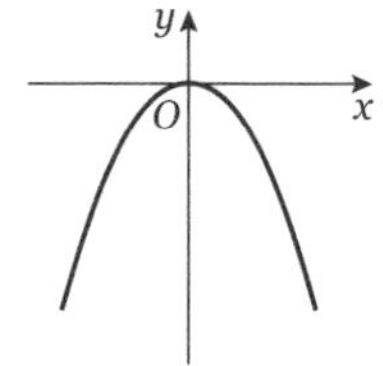

7

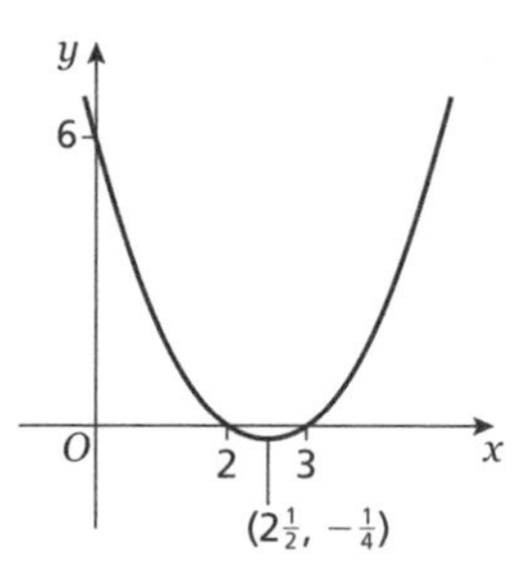

8

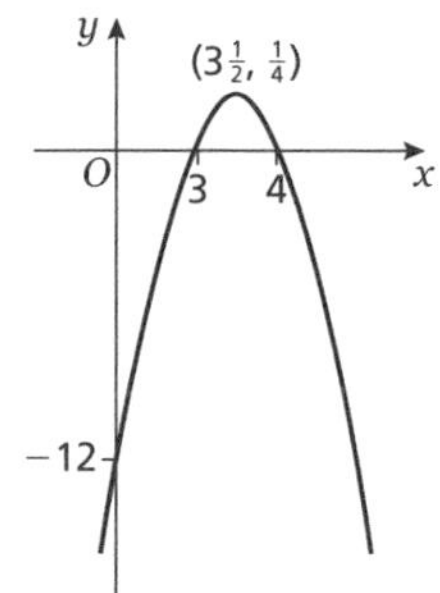

9

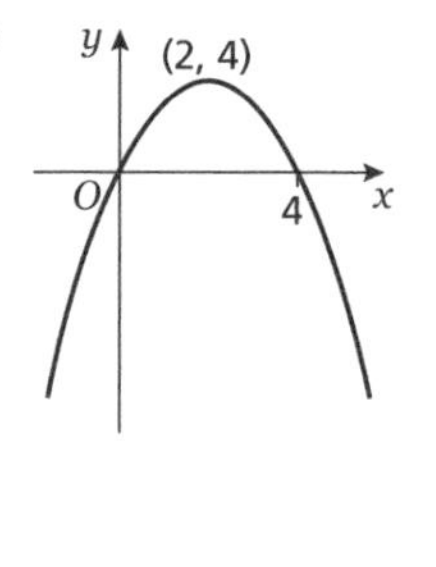

10

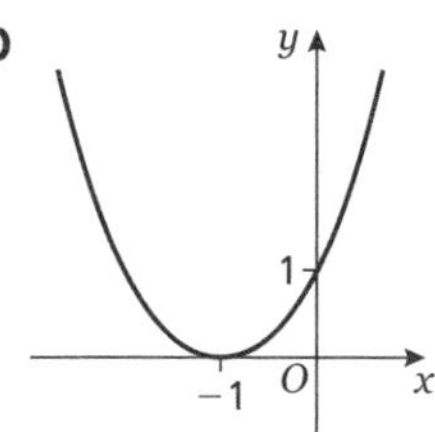

11

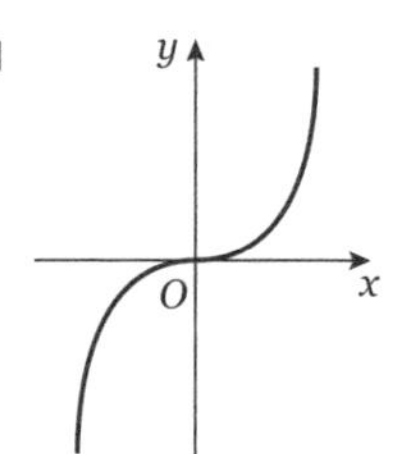

12

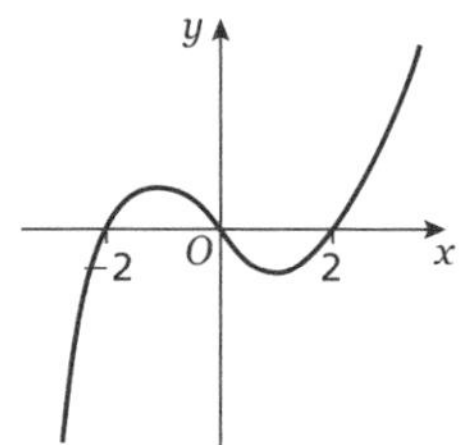

13

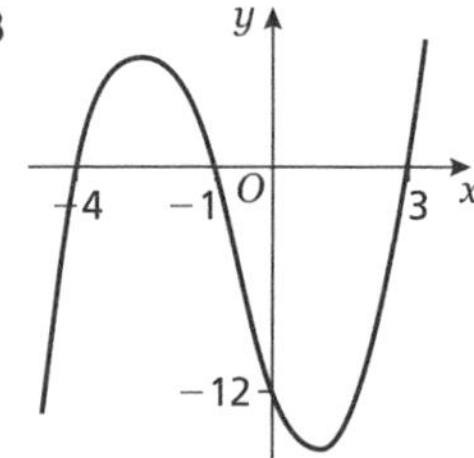

14

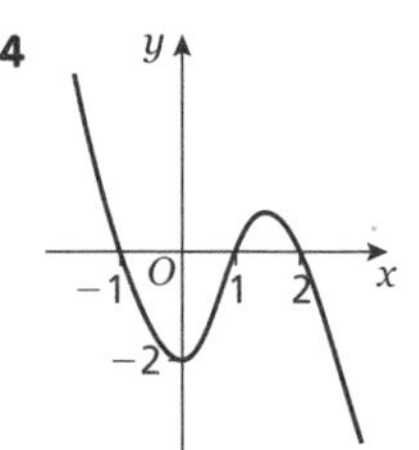

15

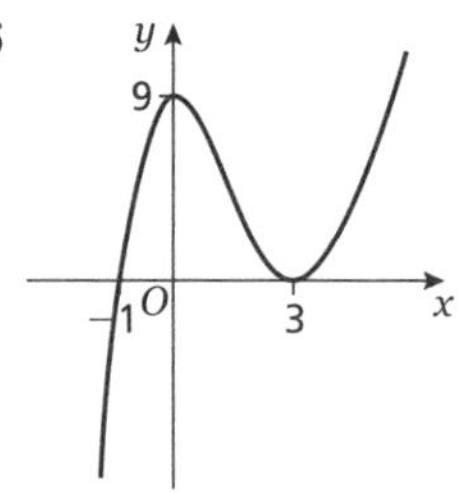

16

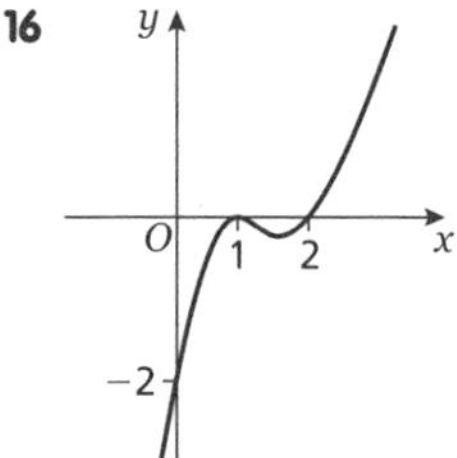

17

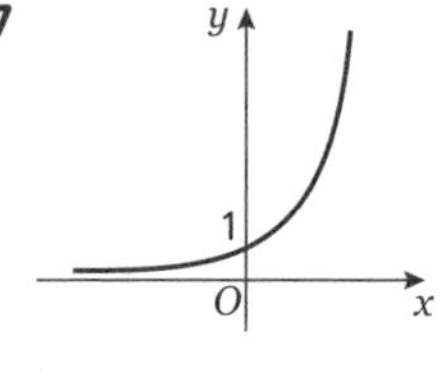

18

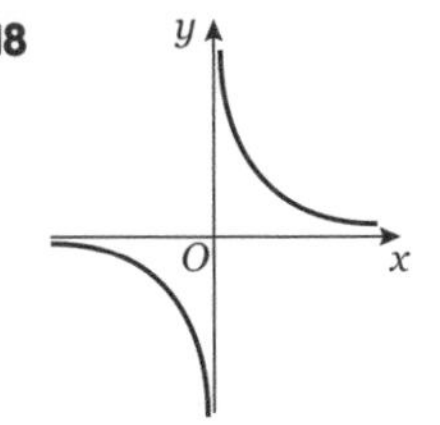

19

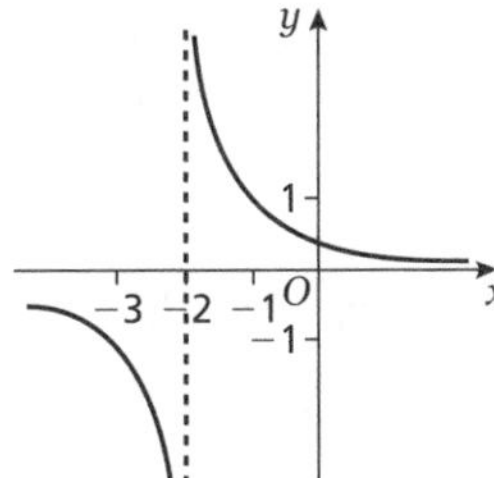

20

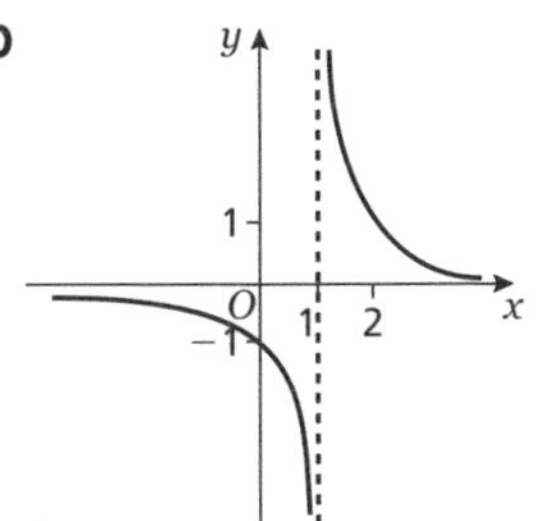

21

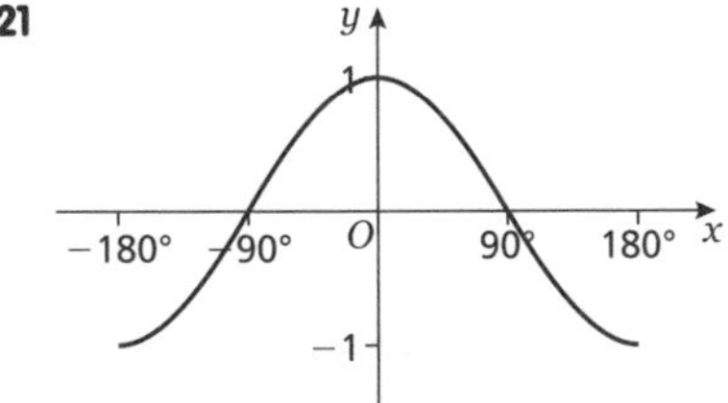

9.4 Graphs of circles

1

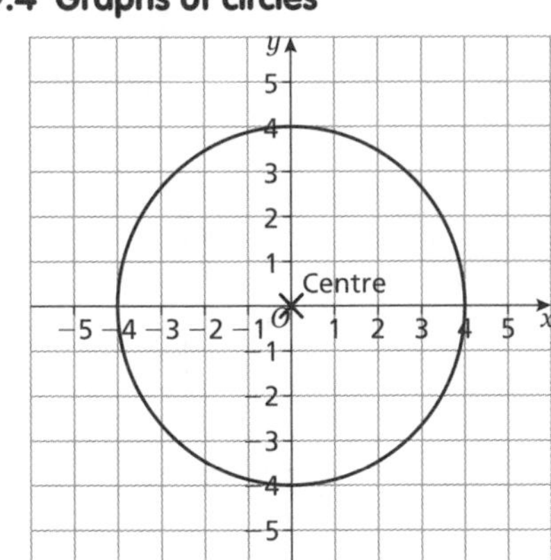

2

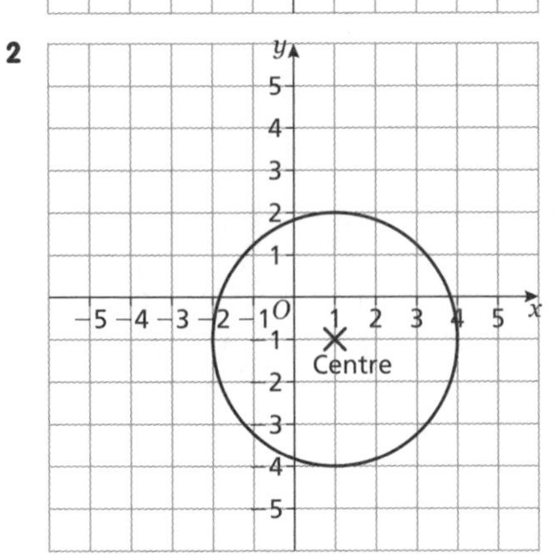

3

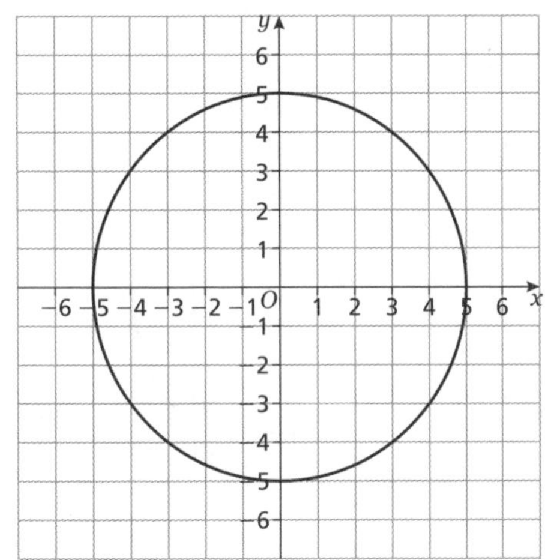

4

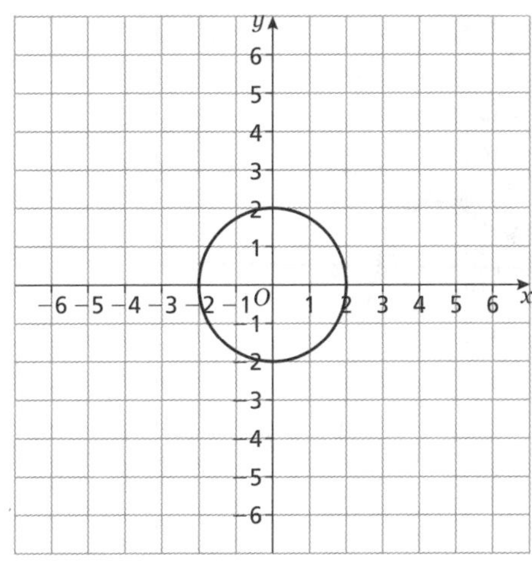

5

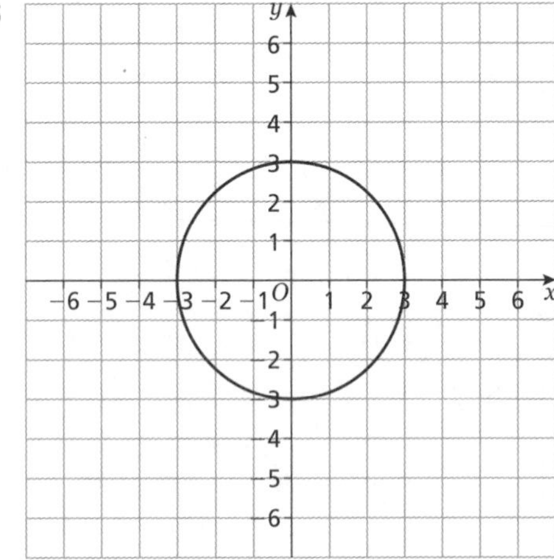

6

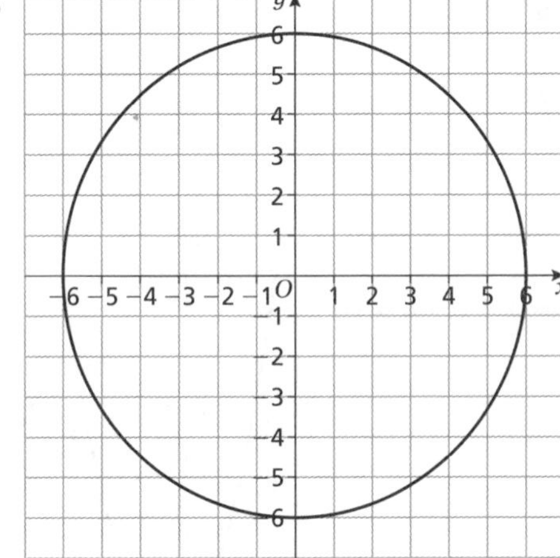

7 **a** $(1, -2)$ **b** 3

c

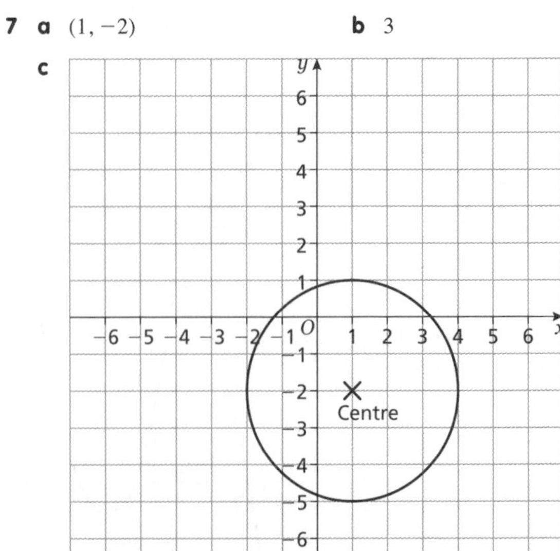

8 **a** $(-2, 3)$ **b** 2

c

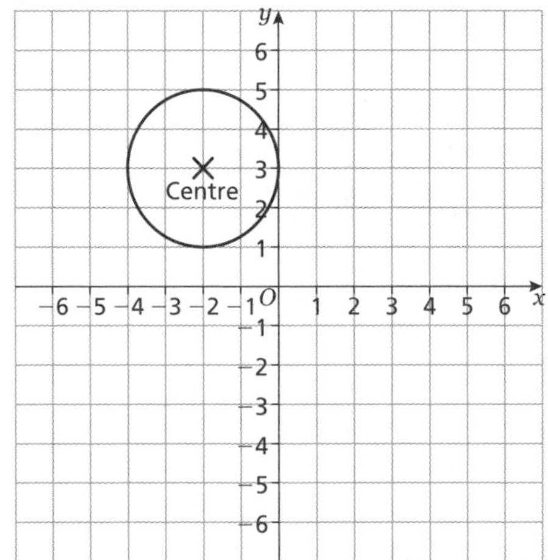

9 **a** $(0, 2)$ **b** 2

c

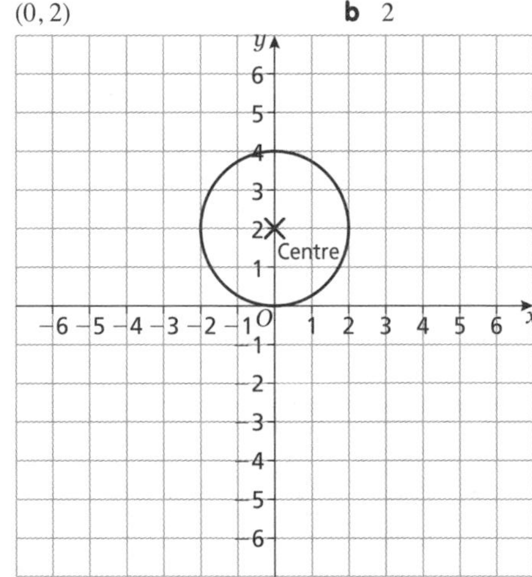

10 $(x - 2)^2 + (y - 3)^2 = 36$

11 centre $= (-2, 5)$, radius $= 4$

Don't forget!

* a straight line
* parabola
*

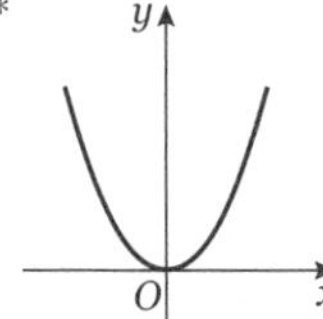

*

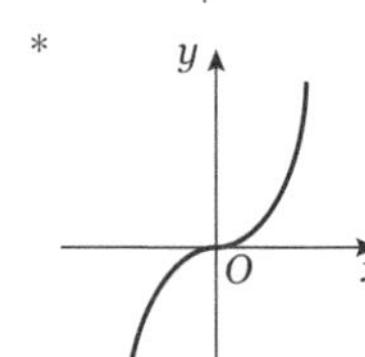

*

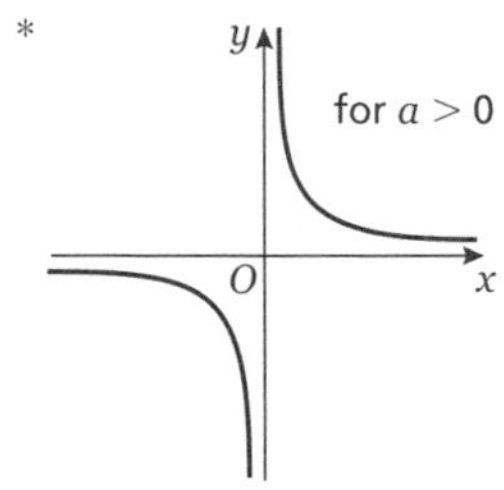

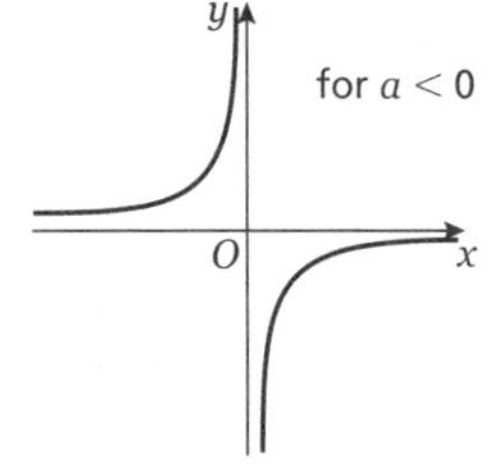

*

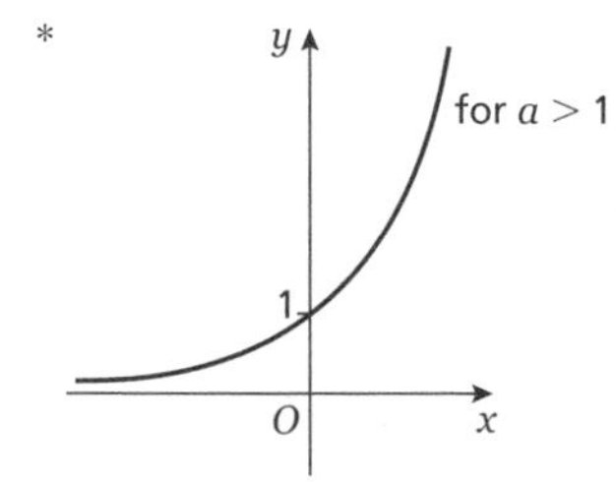

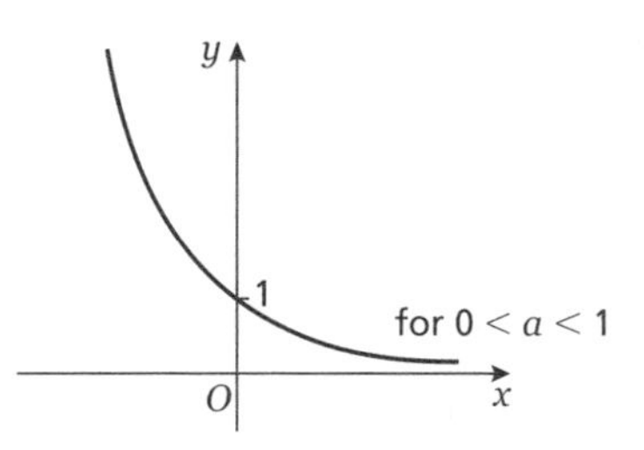

*

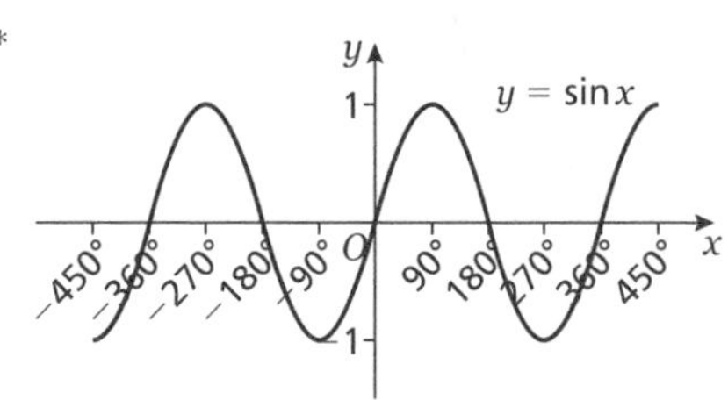

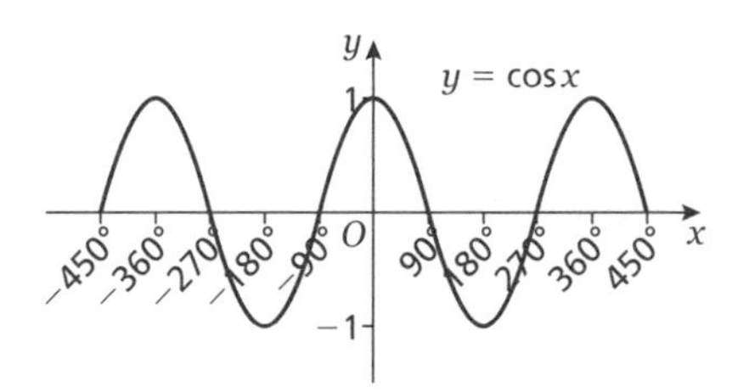

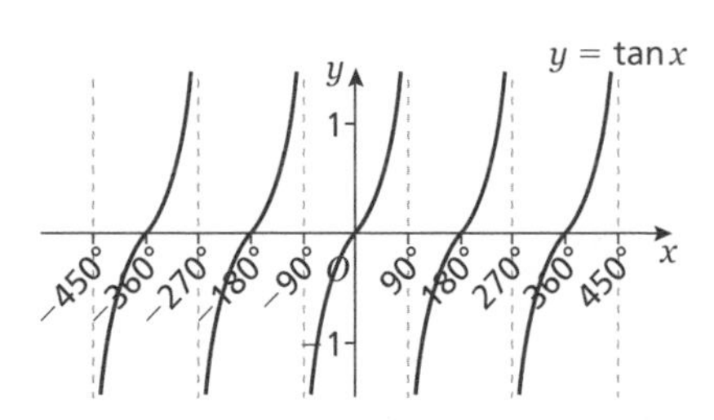

* touches the curve but does not cross it
* perpendicular
* $x = 0$
* $y = 0$
* the curve gets closer to but never touches or crosses
* complete the square
* turning points
* $(x - a)^2 + (y - b)^2 = r^2$; centre; radius; $x^2 + y^2 = r^2$

Exam-style questions

1

Equation	Graph
$y = 3^x$	D
$y = (x + 2)(x - 2)$	C
$y = (2 - x)(2 + x)$	A
$y = \frac{2}{x}$	E
$y = (x + 2)^2(1 - x)$	B

2

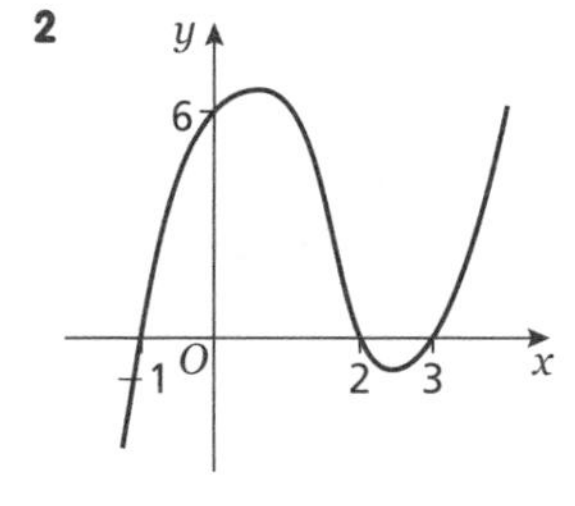

3 **a**

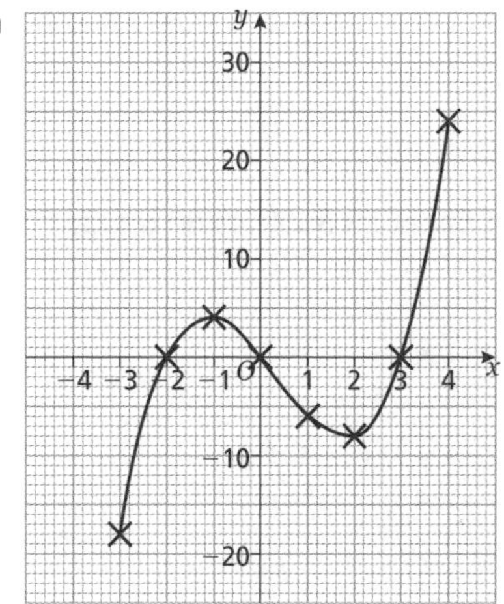

b $-2.4, 0.8, 2.6$

4

10 Inequalities

10.1 Solving linear inequalities

1 **a** $-2 \leqslant x < 4$ **b** $\frac{4}{5} < x \leqslant 2$

c $2x < 12$
$x < 6$

d $-5x \geqslant -10$
$x \leqslant 2$

e $4x - 8 > 27 - 3x$
$7x > 35$
$x > 5$

2 **a** $x \leqslant -4$ **b** $-1 \leqslant x < 5$ **c** $x \leqslant 1$
d $x < -3$ **e** $x > 2$ **f** $x \leqslant -6$

3 **a** $x < -6$ **b** $x < \frac{3}{2}$

4 $x > 5$ (which also satisfies $x > 3$)

10.2 Solving quadratic inequalities

1 $(x + 3)(x + 2) = 0$
$x = -3, x = -2$
$x < -3$ or $x > -2$

2 $x(x - 5) = 0$
$x = 0, x = 5$

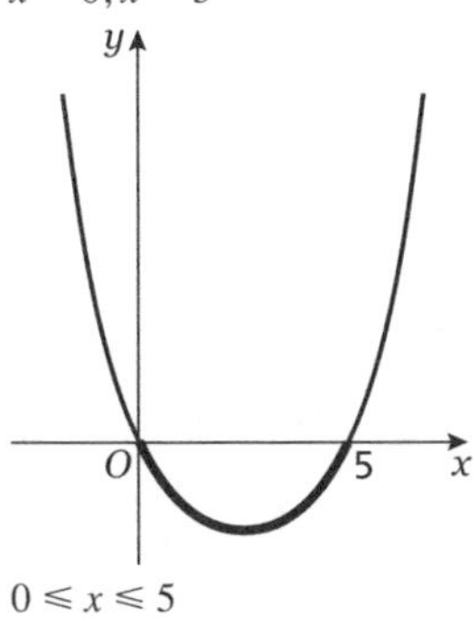

$0 \leqslant x \leqslant 5$

3 $x^2 + 3x - 10 = 0$
$(x + 5)(x - 2) = 0$
$x = -5, x = 2$

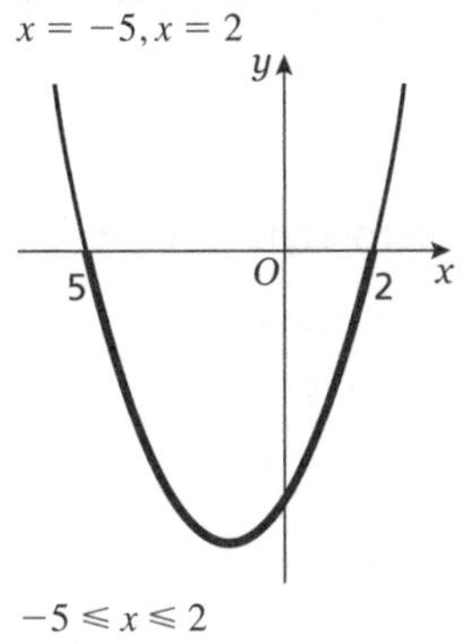

$-5 \leqslant x \leqslant 2$

4 $x \leqslant -2$ or $x \geqslant 6$

5 $-7 \leqslant x \leqslant 4$

6 $\frac{1}{2} < x < 3$

7 $x < -\frac{3}{2}$ or $x > \frac{1}{2}$

8 $-3 \leqslant x \leqslant 4$

9 $2 < x < 2\frac{1}{2}$

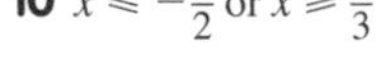

10 $x \leqslant -\frac{3}{2}$ or $x \geqslant \frac{5}{3}$

10.3 Representing linear inequalities on a graph

1

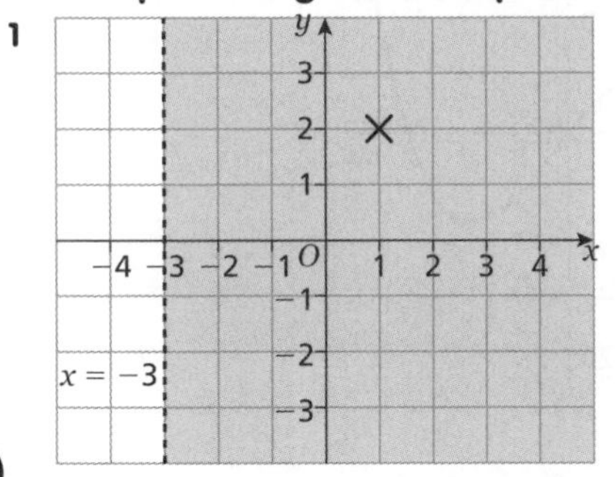

2

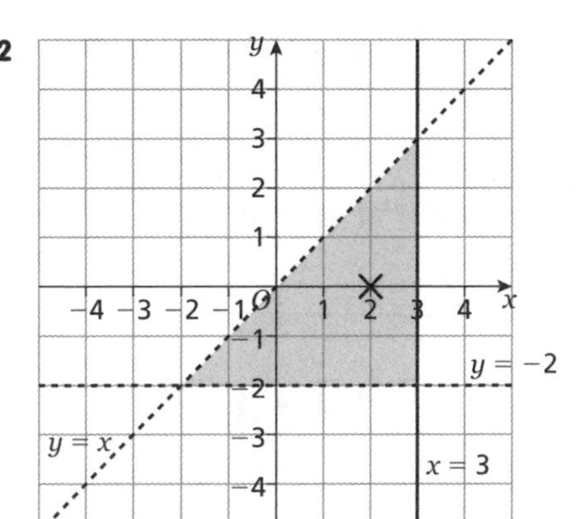

3

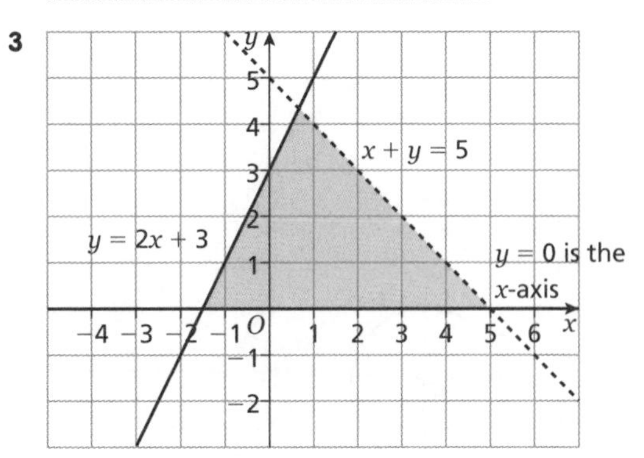

4

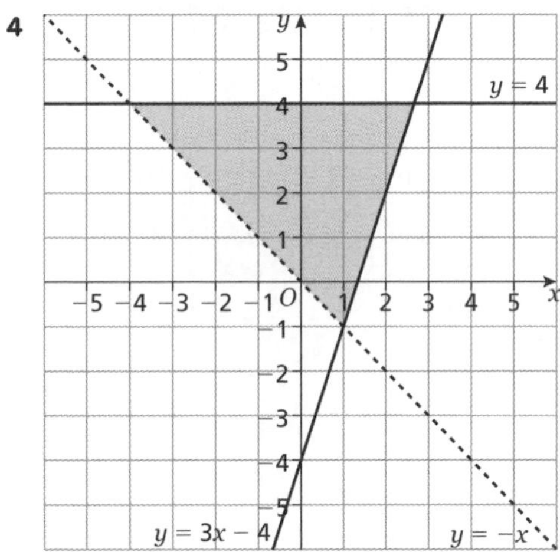

5

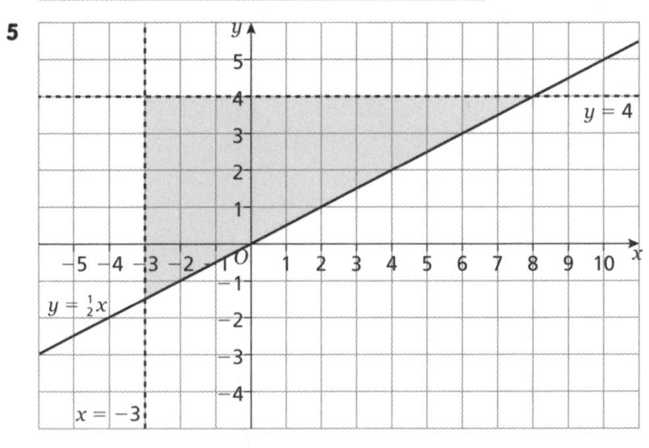

6

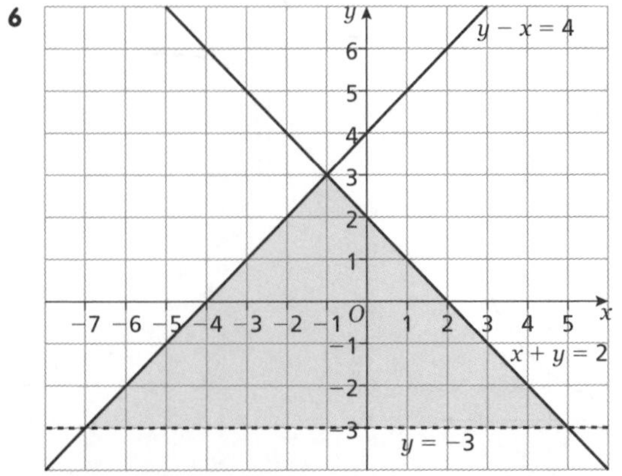

7

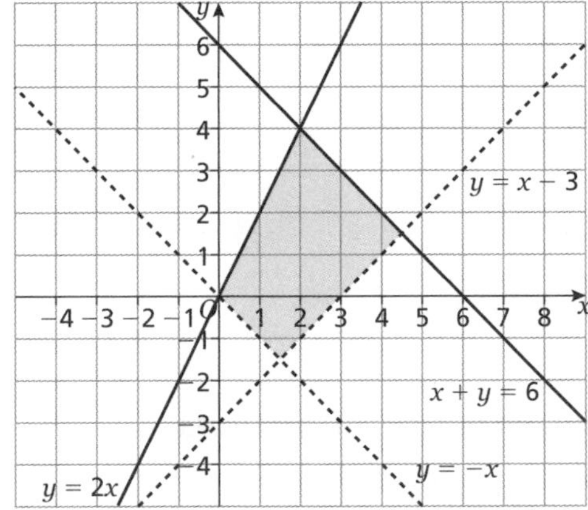

8

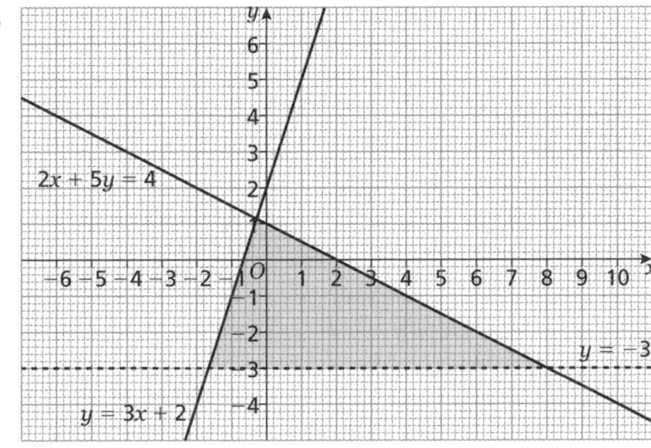

Don't forget!

* a negative number
* solve; sketch the graph; values
* shading regions
* unbroken (solid)
* broken lines

Exam-style questions

1 $-3 \leqslant x \leqslant 2$

2

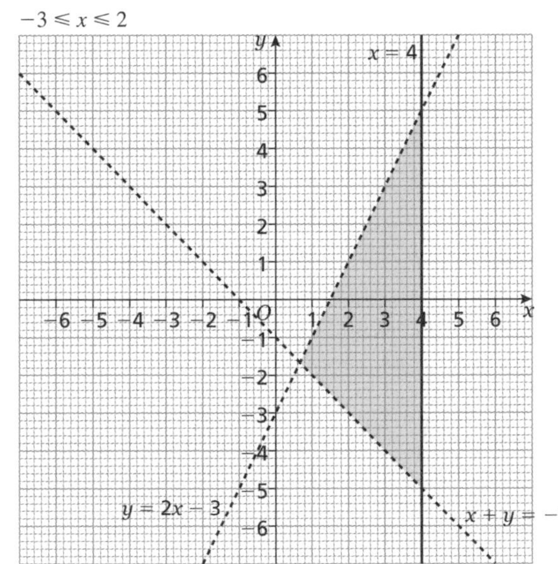

11 Distance–time and speed–time graphs

11.1 Distance–time graphs

1 **a** 2 or 4 (depending on whether you've counted the start and finish)
b $6 \div 10 = 0.6$ m/s
c $21 \div 20 = 1.05$ m/s
d Between 0 and 10 seconds

2 **a** 15 min **b** 15 miles
c 6 mph **d** 48 mph

3 **a** 2 min **b** 10 min
c

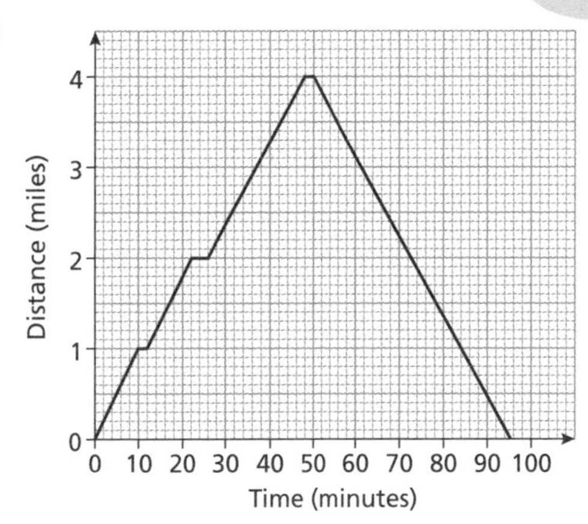

4 **a** C **b** A

11.2 Speed–time graphs

1 **a** $9 \div 12 = 0.75$ m/s^2
b Method 1: $\frac{1}{2} \times 9 \times (20 + 32) = \frac{1}{2} \times 9 \times 52 = 234$ m
Method 2: $\frac{1}{2} \times 9 \times 12 + 9 \times 20 = 54 + 180 = 234$ m

2 **a** $4 \div 8 = 0.5$ m/s^2
b $\frac{1}{2} \times 4(12 + 22) = 68$ m

3 **a**

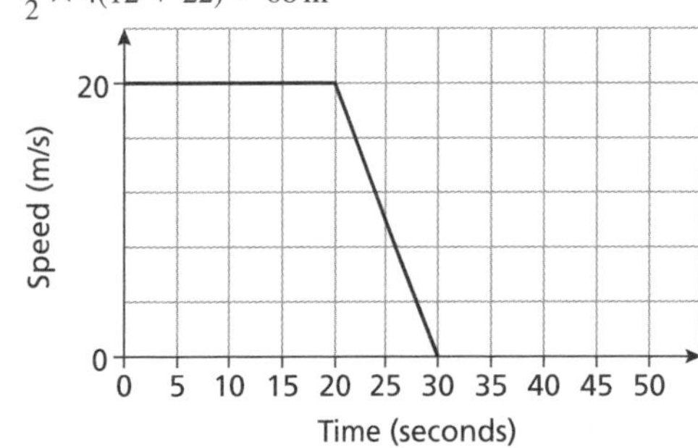

b 2 m/s **c** 100 m

4 **a** 100 km/h^2 **b** 57 miles

5 **a** 4 m/s^2 **b** 150 m **c** -2 m/s^2

6 **a** 30m/s **b** 1 m/s^2

7 **a** 5 s **b** 275 m

Don't forget!

* distance
* the time taken
* speed
* faster
* no movement
* speed
* the time taken to travel
* constant speed
* acceleration
* deceleration
* the distance travelled

Exam-style questions

1 **a** 4 m/s^2
b 150 m

12 Direct and inverse proportion

12.1 Direct proportion

1 **a** $P \propto h$
$P = kh$
$56 = k \times 8$
$k = 56 \div 8 = 7$
$P = 7h$

b $P = 7h$
$P = 7 \times 11$
$P = £77$

2 **a** $y = kx^2$
$45 = k \times 3^2$
$k = 45 \div 9 = 5$
$y = 5x^2$

b $y = 5x^2$
$y = 5 \times 5^2$
$y = 125$

c $y = 5x^2$
$20 = 5 \times x^2$
$x^2 = 20 \div 5 = 4$
$x = 2$

3 a $x = 7y$

b

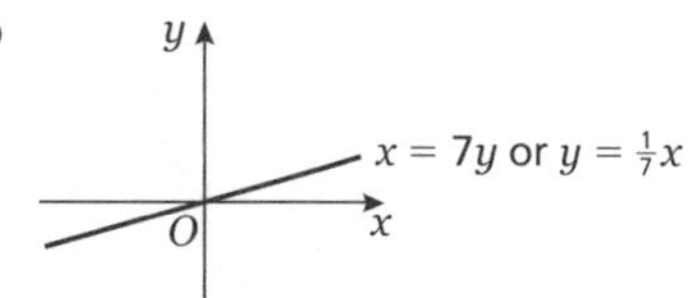

c 91

d 9

4 a $Q = 3Z^2$

b

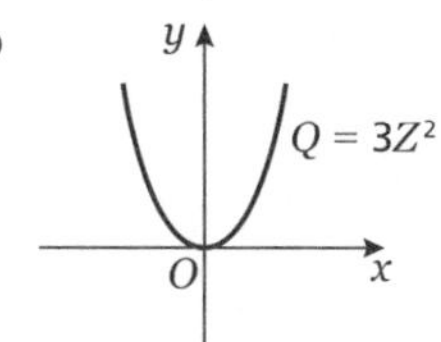

c 75

d 10

5 a $y = 2.5x^2$

b $y = 2.5x^2$

c 6

6 a $B = 2\sqrt{C}$ **b** 16 **c** 100

7 a $C = \frac{2}{3}D$ **b** 300

8 a $x = 3y$ **b** 11.1

9 a $m = 2n^3$ **b** 5

12.2 Inverse proportion

1 a $100 = \frac{k}{10}$

$k = 100 \times 10$

$k = 1000$

$P = \frac{1000}{Q}$

b $P = \frac{1000}{Q}$

$20 = \frac{1000}{Q}$

$Q = 1000 \div 20$

$Q = 50$

2 a $y \propto \frac{1}{\sqrt{x}}$

$y = \frac{k}{\sqrt{x}}$

$1 = \frac{k}{\sqrt{25}}$

$k = 1 \times 5$

$k = 5$

$y = \frac{5}{\sqrt{x}}$

b $y = \frac{5}{\sqrt{x}}$

$5 = \frac{5}{\sqrt{x}}$

$\sqrt{x} = 5 \div 5 = 1$

$x = 1$

3 a $s = \frac{4}{t}$

b $s = \frac{4}{t}$

c 4

4 a $a = \frac{100}{b}$ **b** 2 **c** 10

5 a $v = \frac{80}{w}$

b $v = \frac{80}{w}$

c 40

6 a $L = \frac{36}{W}$ **b** 6

7 a $s = \frac{72}{t}$ **b** 24 **c** 4

8 a $y = \frac{16}{x^2}$ **b** 1

9 a $a = \frac{0.2}{b}$ **b** 0.1 **c** 0.1

Don't forget!

* direct
* inverse
* $\propto$

$\propto x$; $= kx$

$\propto \frac{1}{x}$; $= \frac{k}{x}$

*

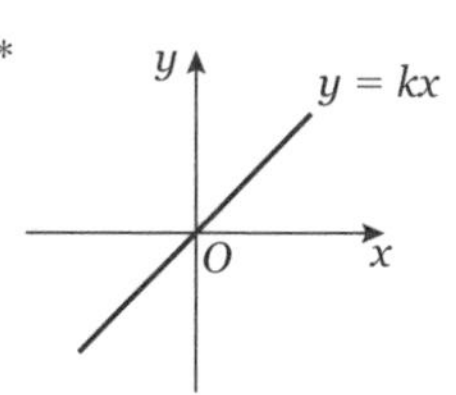

*

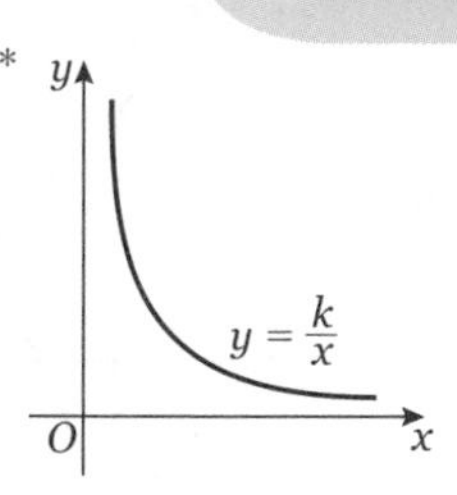

Exam-style questions

1 a $A = 3B^2$ **b** $\frac{3}{4}$ **c** 0.6

13 Transformations of functions

13.1 Applying the transformations $y = f(x) \pm a$ and $y = f(x \pm a)$ to the graph of $y = f(x)$

1

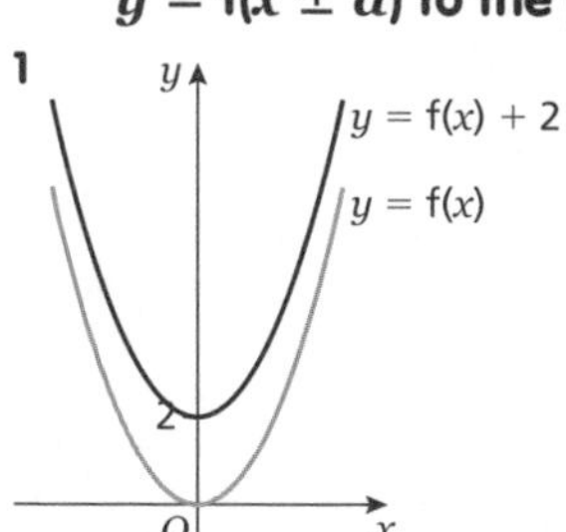

2

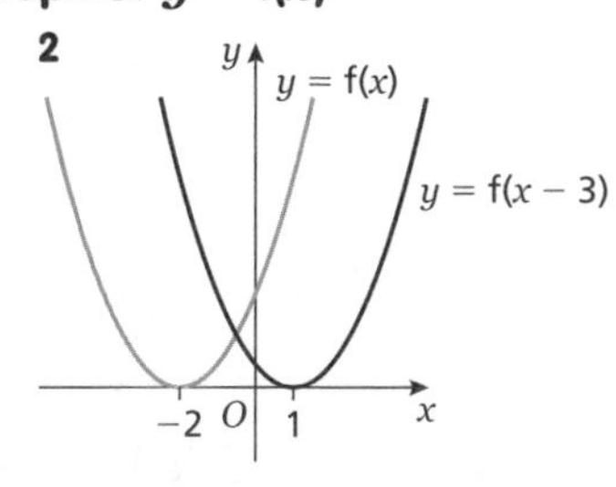

3

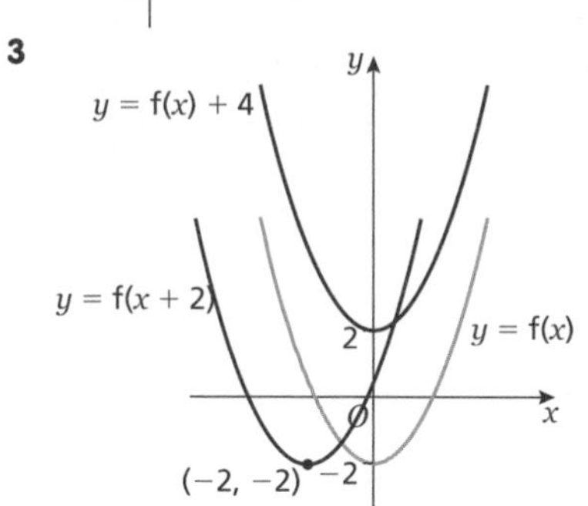

4

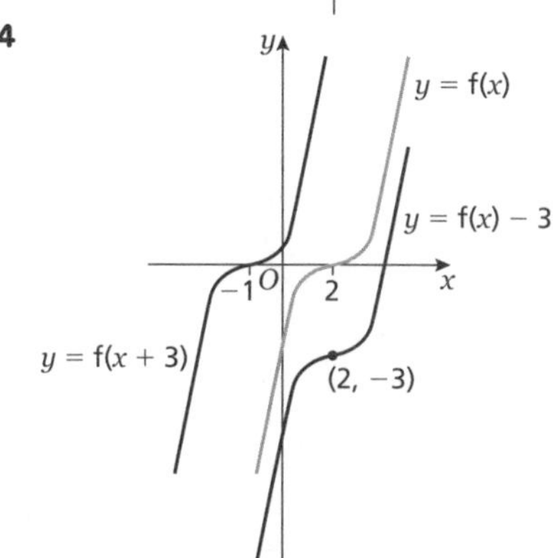

5

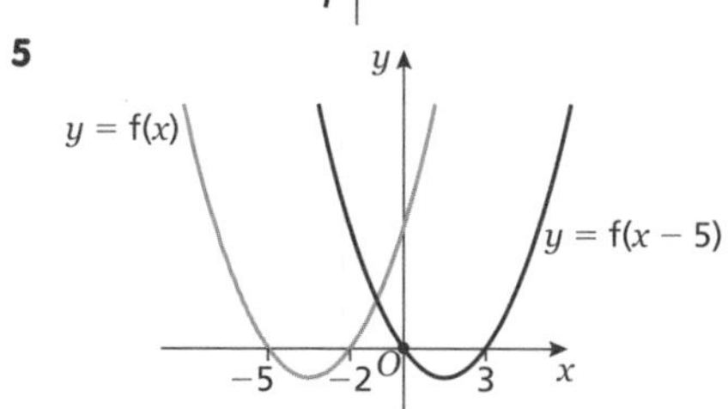

6 C_1: $y = f(x - 90°)$
C_2: $y = f(x) - 2$

7 C_1: $y = f(x - 5)$
C_2: $y = f(x) - 3$

8

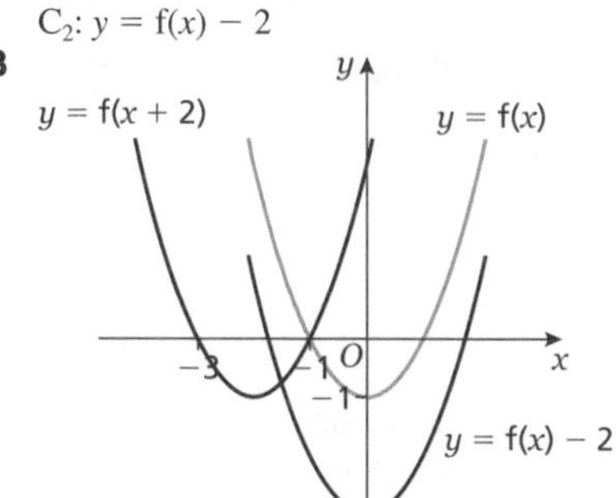

13.2 Applying the transformations $y = f(\pm ax)$ and $y = \pm af(x)$ to the graph of $y = f(x)$

1

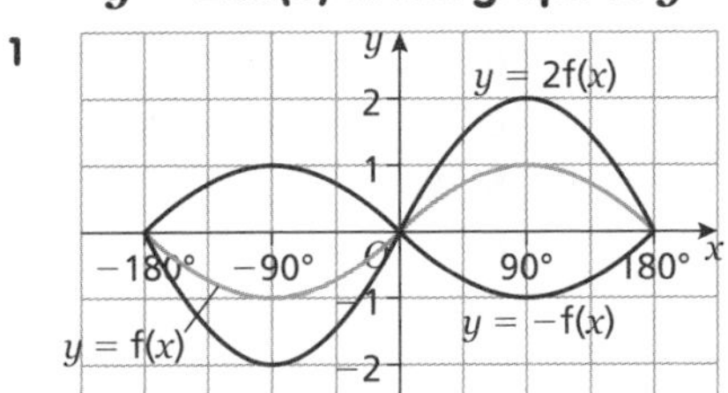

2

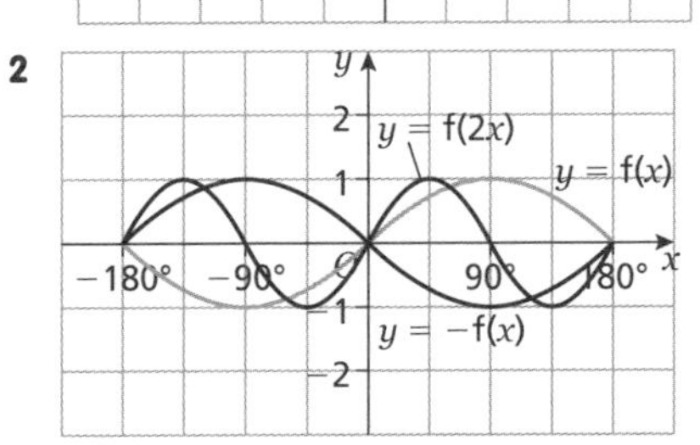

3 a

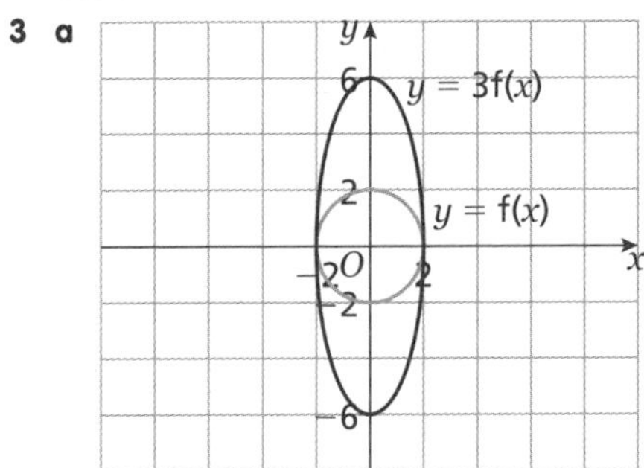

b

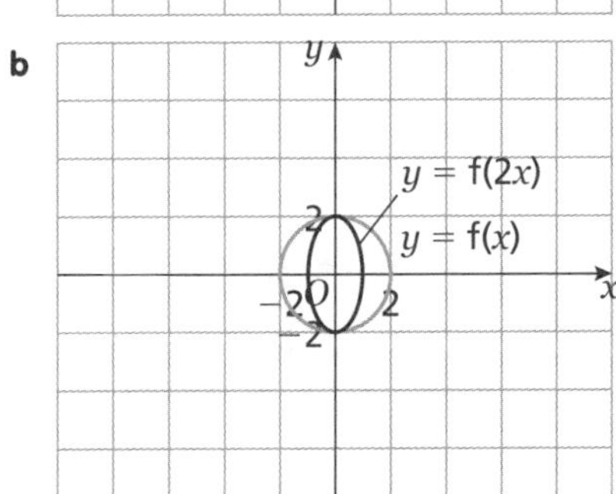

4

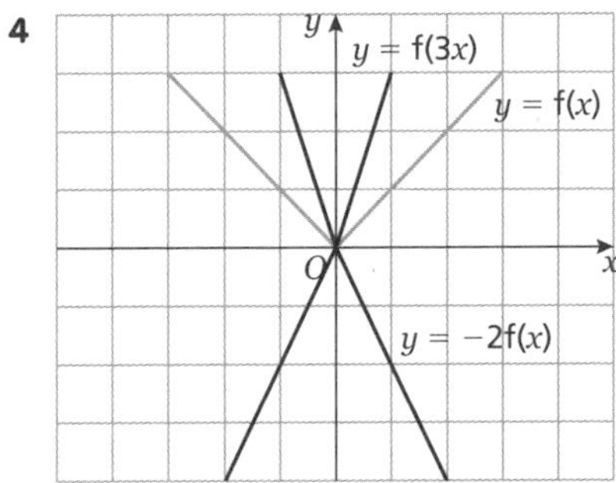

5

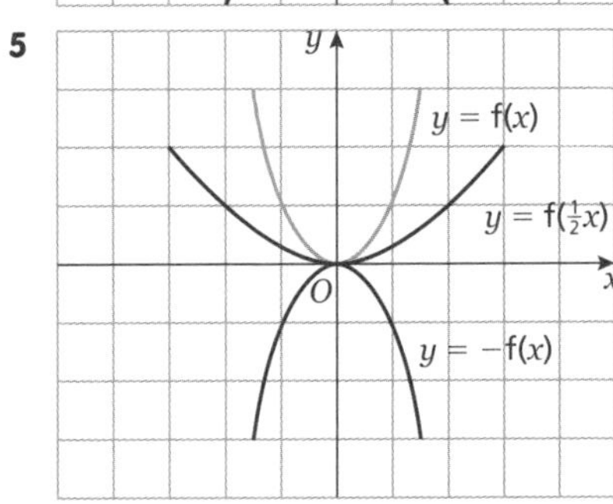

6

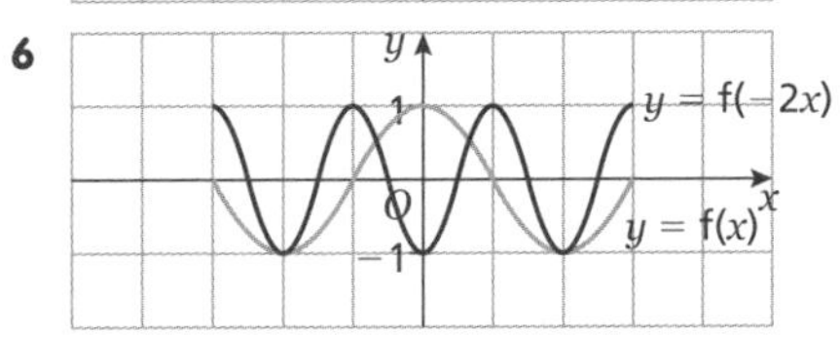

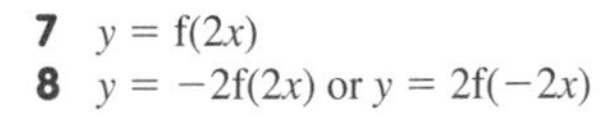

7 $y = f(2x)$

8 $y = -2f(2x)$ or $y = 2f(-2x)$

9 a, b

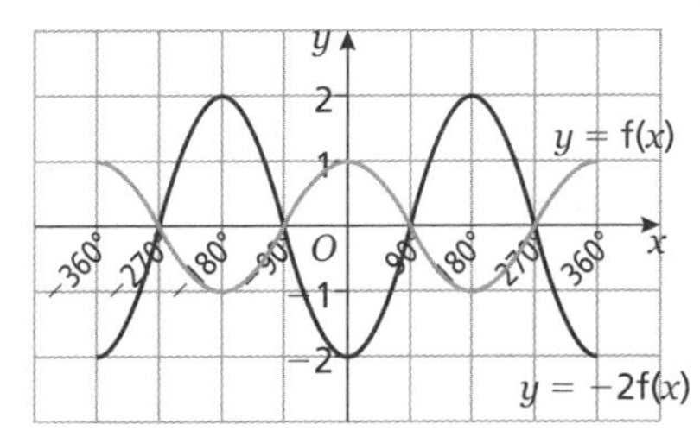

10 a, b

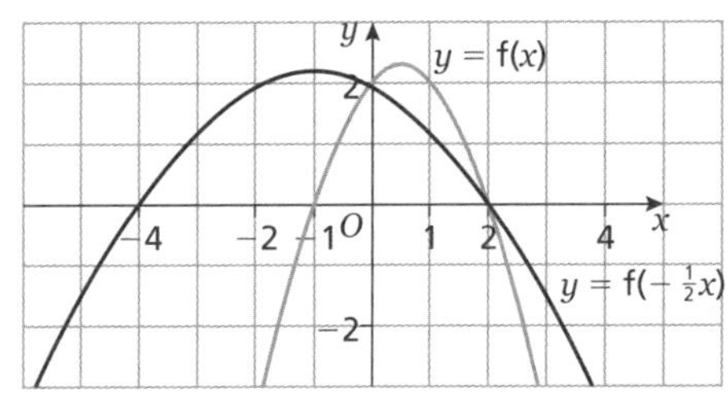

Don't forget!

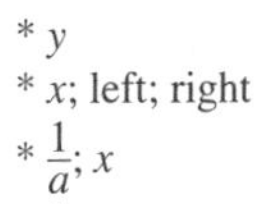

* y
* x; left; right
* $\frac{1}{a}$; x
* $\frac{1}{a}$; x; y
* a; y
* a; y; x

Exam-style questions

1 a

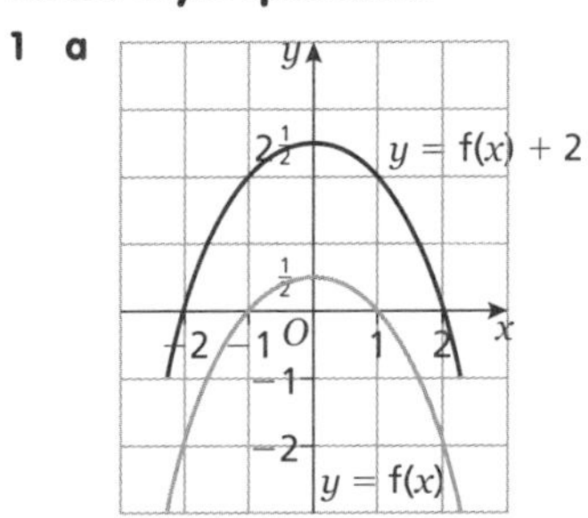

b

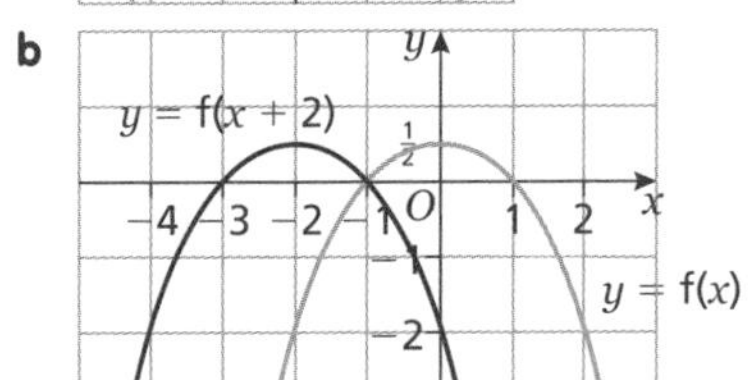

2 a **b**

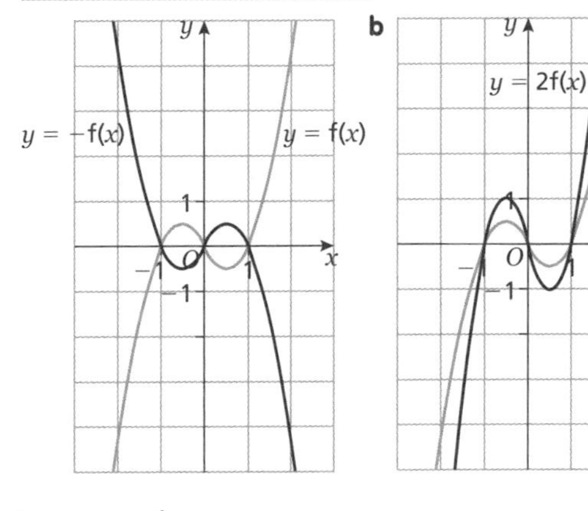

14 Area under a curve

14.1 The trapezium rule

1 $h = 1$

x	0	1	2	3
$y = (3 - x)(x + 2)$	6	6	4	0

$y_0 = 6, y_1 = 6, y_2 = 4, y_3 = 0$

$A = \frac{1}{2} \times 1 \times [6 + 2(6 + 4) + 0]$

$= \frac{1}{2}[26]$

$= 13$ sq units

2 $h = \frac{10-4}{3} = 2$

x	4	6	8	10
y coordinate for the curve	7	12	13	4
y coordinate for the straight line	7	6	5	4

$y_0 = 0, y_1 = 6, y_2 = 8, y_3 = 0$

$A = \frac{1}{2} \times 2\,[0 + 2(6 + 8) + 0]$

$= 1 \times 28$

$= 28$ sq units

3 34 sq units **4** 149 sq units **5** 14 sq units

6 $25\frac{1}{4}$ sq units **7** 35 sq units **8** 42 sq units

9 $26\frac{7}{8}$ sq units **10** 56 sq units **11** $6\frac{1}{4}$ sq units

Don't forget!

* the area under a curve

* Area $= \frac{1}{2}h[y_0 + 2(y_1 + y_2 \ldots + y_{n-1}) + y_n]$; the values of y for each value of x used

* number of equal strips the area has been divided up into; the vertical boundaries of the area

* the number of strips, n

* $= \frac{b-a}{n}$

Exam-style questions

1 71.25 sq units

2 35 sq units

3 72 sq units

Practice Paper

1 $x = -4, y = -5$ **2** $m = \pm\sqrt{\frac{k}{6}}$

3 $4y^2 - 6xy - 6x^2$

4 a $(2x - 3)(x + 1)$ **b** 525

5 a

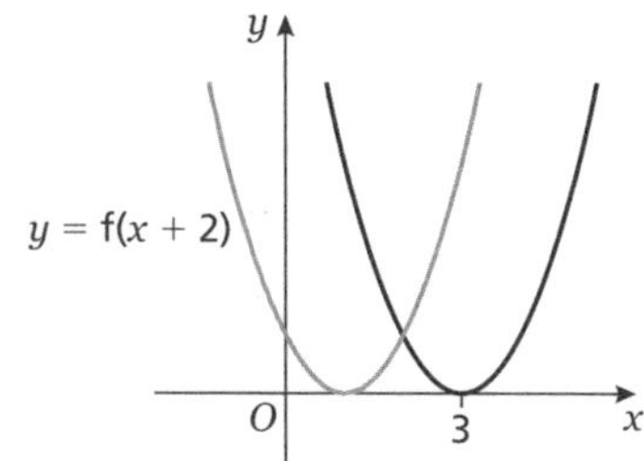

Minimum point at (1, 0)

b

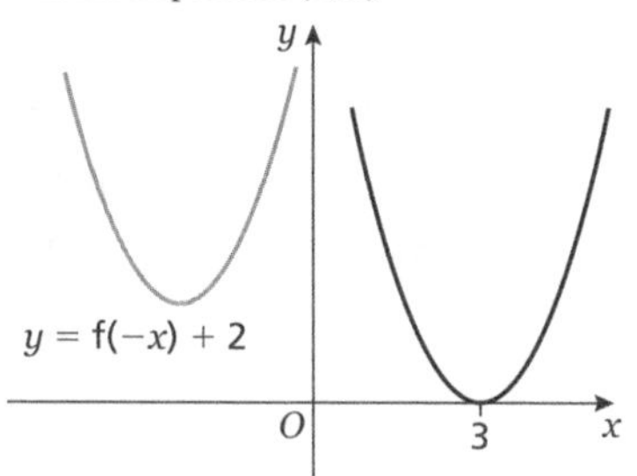

Minimum point at (−3, 2)

6 a $3 \pm \sqrt{2}$ **b** $45 - 29\sqrt{2}$

7 a $\frac{1}{2x}$ **b** $\frac{c}{9a^4b^2}$ **c** $\frac{1}{3}$

8 a $b = 1, c = -6$ **b** $a = 4, q = 16$

9 a $x > \frac{5}{3}$ **b** $-3, -2, -1, 0, 1, 2, 3$

10 a 399 **b** 22

11 a $\frac{4-1}{6-2} \times \frac{8-4}{3-6} = -1$ **b** 12.5

12 a

x	−3	−2	−1	0	1	2	3	4
y	**−18**	0	**4**	0	**−6**	−8	**0**	24

b

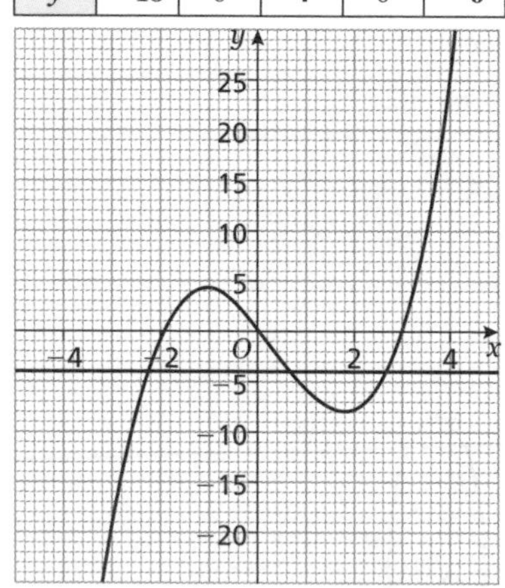

c x is approximately −2.3, 0.6 or 2.7

13 a $\frac{4x-3}{x^2-1}$ **b** $\frac{3}{4}$

14 a $16p^2 - 4 \times 4 \times (4p + 5) = 16p^2 - 64p - 80$

b −1, 5

15 a

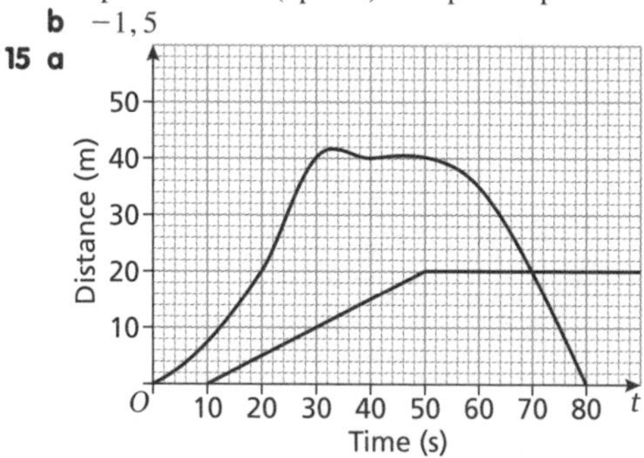

b 70 **c** 1.7 ± 0.2 m/s

16 a (3, −2) **b** A (3, 2); B (3, −6) **c** $-2 \pm 2\sqrt{3}$

17 a 540 g

b $C \propto d$

$C = kd$

$36 = k \times 12$

$C = 3d$

$= 3\sqrt{\frac{3m}{5}}$

$= 3 \times \sqrt{\frac{9m}{15}}$

$= 9\sqrt{\frac{m}{15}}$

18 a (4, 12) **b** 14 sq units